Laser in der Umweltmeßtechnik
Laser in Remote Sensing

Vorträge des 11. Internationalen Kongresses
Proceedings of the 11th International Congress

Laser 93

Herausgegeben von/Edited by
C. Werner, W. Waidelich

Mit 130 Abbildungen/With 130 Figures

Springer-Verlag Berlin Heidelberg GmbH

Dr. rer. nat. Christian Werner

Deutsche Forschungsanstalt für Luft- und Raumfahrt,
Institut für Optoelektronik, Oberpfaffenhofen

Dr. rer. nat. Wilhelm Waidelich

Universitätsprofessor, em. Vorstand des Instituts
für Medizinische Optik der Universität München, em. Direktor
des Instituts für Angewandte Optik der Gesellschaft
für Strahlen- und Umweltforschung , Neuherberg

ISBN 978-3-540-57443-9

Die Deutsche Bibliothek - CIP-Einheitsaufnahme
Laser in der Umweltmeßtechnik: Vorträge des 11.Internationalen Kongresses Laser 93 =
Laser in remote sensing/ hrsg. von C. Werner; W. Waidelich.

ISBN 978-3-540-57443-9   ISBN 978-3-662-08252-2 (eBook)
DOI 10.1007/978-3-662-08252-2
NE: Werner, Christian (Hrsg.); Internationaler Kongress Laser -11, 1993, München -; PT

Satz: Reproduktionsfertige Vorlagen vom Autor

62/3020 - 5 4 3 2 1 - Gedruckt auf säurefreiem Papier

Vorwort

Mit der Kongress-Messe LASER 93 eröffnete die Münchener Messe- und Ausstellungsgesellschaft einen Überblick über den neuesten Stand und die Entwicklungstendenzen. Die Synthese von Forschung und Anwendung (Kongress) mit der Praxis der kommerziell erhältlichen Geräte (Messe), wie sie im Verbund von Kongress und Ausstellung dokumentiert wird, hat auf der LASER langjährige Tradition.

In den letzten Jahren hat sich die Optoelektronik und die Anwendung des Lasers mit einer solchen Dynamik weiterentwickelt, daß es sinnvoll wurde, die Beiträge des Kongresses bereits anläßlich der LASER 91 in nach Anwendungen getrennten Bänden zu dokumentieren.

Der hier vorliegende Band *Laser in der Umweltmeßtechnik* zeigt die Bedeutung der optischen Meßtechnik für das wachsende Gebiet der Umweltforschung. Die Organisation des Teilkongresses wurde auf mehrere Spezialisten verteilt.

Die Kommission Reinhaltung der Luft im VDI und DIN hat ein Arbeitsgruppentreffen abgehalten (K.Weber), um Richtlinien für den Einsatz optischer Fernmeßverfahren zu erarbeiten. Dies geht auf Anregungen zurück, die anläßlich des Workshops auf der LASER 91 erarbeitet wurden. Ein Glossar für die Fernerkundung ist Bestandteil der Richtlinienarbeit und ist im vorliegenden Band als Entwurf enthalten (V.Klein).

Die meisten Beiträge wurden in Englisch vorgelegt, vier Beiträge, die während der Konferenz präsentiert wurden, liegen schriftlich nicht vor.

Für das Zustandekommen des Buches sei allen Autoren, Sitzungsleitern und Diskussionsteilnehmern und dem Springer Verlag gedankt.

München, August 1993

Christian Werner
Wilhelm Waidelich

Preface

Since 1973, the Munich Trade Fair Corporation has been providing a survey of the state-of-the-art and trends in development through its LASER OPTOELECTRONICS congress and trade fair. The successful synthesis of theory and practice made evident by the coordination of the congress and exhibition, has helped this high-tech information forum in Munich to gain an international reputation. This is the oldest and most important event of this kind and it has become the meeting place for experts from all over the world.
During the last years the development of optoelectronics and laser applications was increasing at such a rate, that it would be necessary to publish the contributions presented at LASER 91 in separate volumes. Beginning with LASER 91 **Lasers in** Environmental **Remote Sensing** is presented in a separate volume.

This volume **Laser 1993 in (environmental) Remote Sensing** demonstrates the importance of optical measuring techniques for its increasing application in environmental sciences. The organization of the sessions was appointed to different experts

The Commission on Air Pollution Prevention in VDI and DIN held a working group meeting (K.Weber) to elaborate guidelines for optical remote sensing. After the workshop at LASER 91, where experts from industry and research discussed the problem of remote sensing methods for air pollution monitoring, working on guidelines for remote sensing instruments is the next step. A glossary for remote sensing was discussed (V.Klein). We hope for LASER 95 to present the first experiences with the guidelines.

Most of the papers are written in English, four papers presented during the congress are missing in this volume, papers accepted but not presented during the congress are not included.

We would like to express our gratitude to all: Authors, workshop participants, chairpersons, and to Springer-Verlag for their help in the preparation of this book.

Munich, August 1993   Christian Werner
                                                             Wilhelm Waidelich

Referenten - Contributors

Sitzungsleiter - Session Chairmen

M. Tacke	Analysis of Combustion
V. Klein	Remote Sensing of Trace Gases
K. Günther	Vegetation Stress
G. Cecchi	Water Pollution
C. Weitkamp	DLR Lidar Development and Application for Remote Sensing
Ch.Werner	Visibility
K. Weber	Meeting of the VDI Working Group for Optical Remote Sensing; Guidelines for the Use of Lidar Methods

Inhaltsverzeichnis - Contents

3. Vegetation Stress

4. Water Pollution

6.   LIDAR-Visibility Measurements

7.   VDI/DIN Working Group

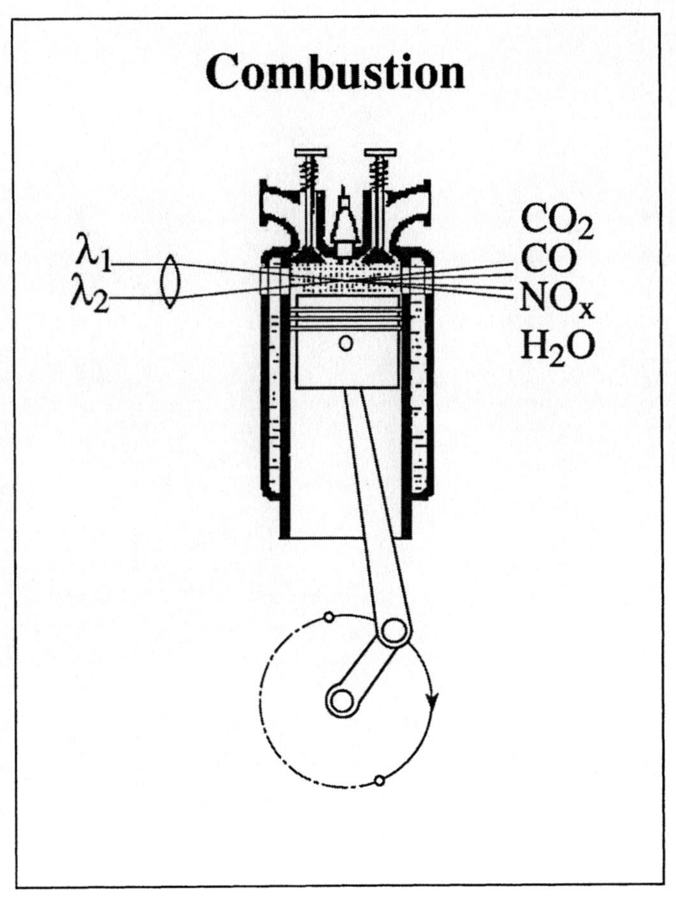

Combustion
λ_1
λ_2
CO_2
CO
NO_x
H_2O

Two-Dimensional Laser Diagnostics for Technical Combustion

Alfred Leipertz

Lehrstuhl für Technische Thermodynamik (LTT-Erlangen)

Universität Erlangen - Nürnberg

Am Weichselgarten 9, D-91058 Erlangen

Fed. Rep. Germany

1. Introduction

Real combustion processes are turbulent and thus three-dimensional in nature revealing large- and small-scale structures in the flow and combustion field. For modeling purposes and thus for a better understanding of the phenomena, quantitative analysis is necessary. This can be done by using point-measurement techniques applying statistical evaluation techniques. Two- and three-dimensional structures, however, cannot be resolved easily. Extending these techniques with quantitative potentials to two- (2D) and three-dimensional (3D) resolving techniques by forming a light sheet and detecting the signal intensity within this sheet by a two-dimensional photo-electrical detector, the required field information can be obtained for a large number of fundamental applications, but also for a few more technical situations, see e.g. LEIPERTZ (1989). Well-established quantitatively working laser techniques providing high spatial and temporal resolution are laser-induced fluorescence (LIF) and the laser scattering techniques (Mie, Rayleigh, Raman). These will be treated here as being used for the investigation of technical combustion. The principles of the different techniques cannot be given in this paper, see e.g., ECKBRETH (1987). Their application to the investigation of technical combustion is presented as they are used in different fields at LTT-Erlangen.

2. Spray diagnostics by laser Mie scattering

Particle Mie scattering is the strongest scattering process applied. It is well known from being used in laser Doppler velocimetry (see e.g., DURST et al. (1976)) and particle size measurements (see e. g., DURST et al. (1985)).
In a two-dimensional (2D) extension it first has been used for imaging the gas concentration in a nozzle flow by LONG et al. (1981). In the meanwhile its application has become widespread for the investigation of technical flows and

flames. We have employed it to the investigation of the injection process in a real VW Diesel engine with optical access providing information on spray penetration and fuel distribution inside the piston bowl detecting the 2D-Mie signal from the Diesel droplets (MÜNCH and LEIPERTZ (1992)).

Figure 1 displays the typical 2D measurement image as taken by the intensified CCD camera by means of a mirror and a window in the bottom of the piston. It contains the information of two different scattering processes resulting in two different signals on the 2D pictures. One of both is caused by a three-dimensional integral scattering signal from droplets of the whole jets inside the piston bowl (bright regions in Fig. 1) generated by droplets outside the irradiated 2D-light sheet plane being illuminated by the stray light from reflections at the engine windows and components. This signal displays the projection of the 3D contour of all jets in observation directions and gives information on the spray tip penetration and cone angle of the individual jets. The other signal part is generated by droplets staying inside the light sheet plane during the shutter time of the camera (dark regions inside the jets in Fig. 1) giving information on the local and temporal distribution of the liquid fuel at that downstream positions inside the piston bowl. Both signals can be used in a different way to characterize the injection process. As an example, Figure 2 displays the results of the tip penetration in dependence on the injection time for the five different jets of the Diesel engine.

3. Combustion field measurements in a gas turbine combustor

Gas combustion diagnostics in these days is mainly directed to the detection of reaction zones, high temperature spots where most NOx is generated and gaseous pollutans being generated at particular local positions inside the combustion process. 2D measurement techniques can provide detailed information on these quantities inside an extended 2D part of the combustion field under investigation within the temporal resolution of a single laser shot. We have used the 2D laser Rayleigh technique for temperature measurements and 2D laser induced fluorescence (2D-LIF) for OH detection indicating the reaction zone. Details on both techniques cannot be given here.

NOx-generation strongly depends on the local instantaneous gas temperature in the flame. Using mechanical probe sensors for gas temperature determintion, e. g. thermocouples, the flow and temperature field may be disturbed to some

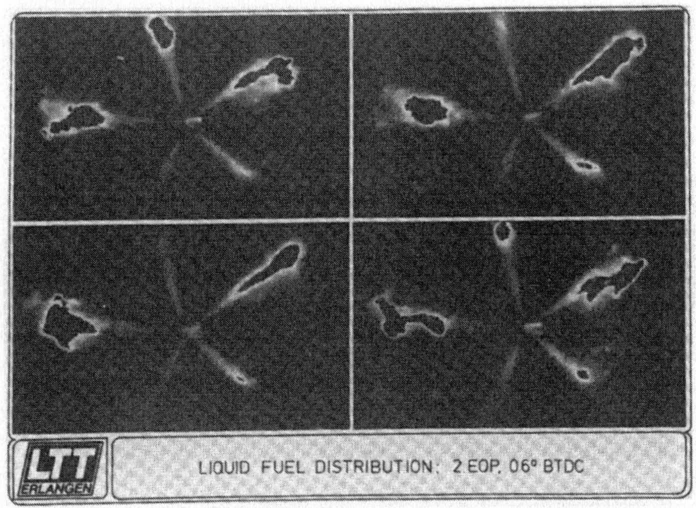

Figure 1: Typical Mie measurement signal inside the piston bowl of a DI Diesel engine at the same injection time of different engine cycles.

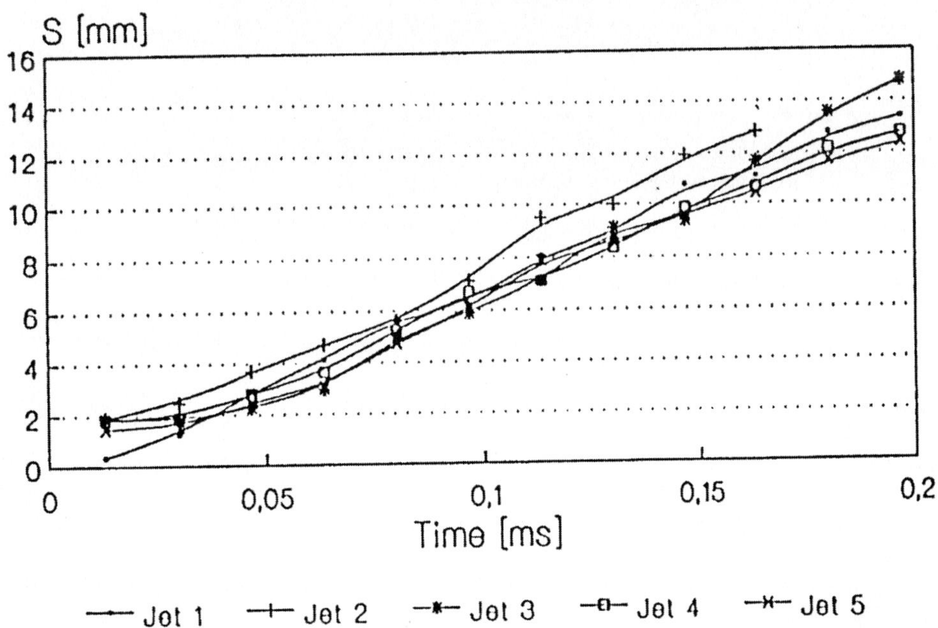

Figure 2: Tip penetration of the five different jets in dependence of the injection time.

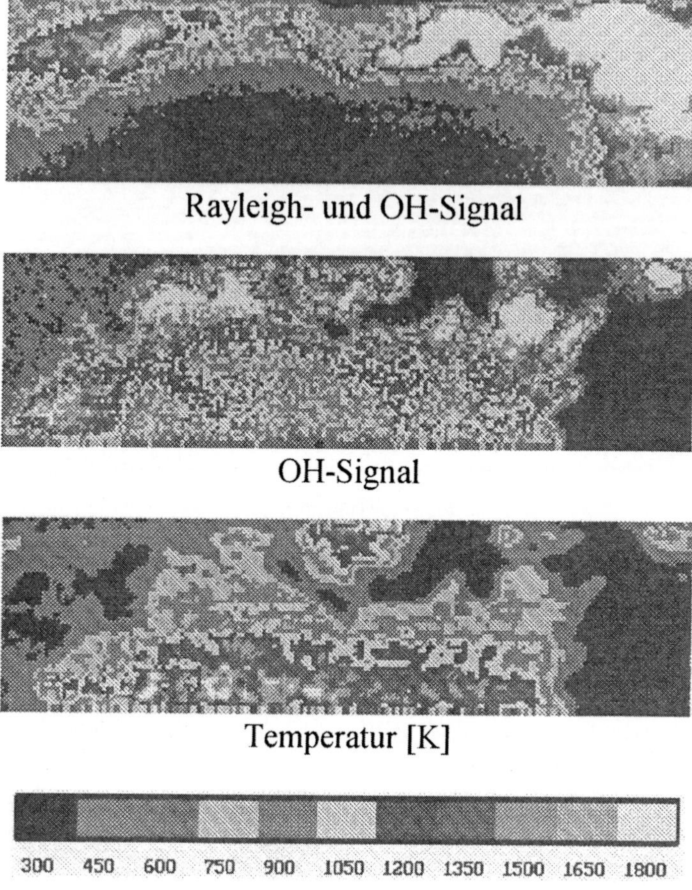

Rayleigh- und OH-Signal

OH-Signal

Temperatur [K]

300 450 600 750 900 1050 1200 1350 1500 1650 1800

Figure 3: Simultaneous detection of 2D Rayleigh and OH-LIF. The temperature values can be deduced from the Rayleigh data.

extend and radiation heat loss must be taken into account appropiately. Laser based measurement techniques, in general, do not influence the flow and combustion field under investigation. For clean environments the laser Rayleigh technique has been proved successfully in small-scale laboratory combustion systems providing point resolved measurement information (DIBBLE and HOLLENBACH (1981), HAUMANN and LEIPERTZ (1984)) and 2D temperature imaging (LONG et al (1985), LEIPERTZ et al (1992) as well. For the first time point resolved temperature measurements in a technical combustor have been performed by BARAT et al (1991).

Very recently, the first application of the 2D laser Rayleigh technique to the investigation of a large-scale industrial combustor has been reported by KAMPMANN et al (1993). Using 2D laser Rayleigh scattering, quantitative measurements of the temperature fields have been performed in different downstream positions of a 150 kW industrial, premixed, turbulent low-emission swirl combustor. Due to the possible interferences of the Rayleigh signal with Mie scattering and laser reflections of the burner components, some minor modifications of the design of the combustor and its gas supply were necessary. This was done without changing the basic characteristics of the burner. The quantitative and instantaneous character of the collected data allowed the calculation of ensemble averaged temperature distributions and analysis of the flame structure in the turbulent combustion field. The measured temperature distribution confirmed that the investigated flame is stabilized by a central recirculation zone.

In the meanwhile this technique has been extended to simultaneously detecting the 2D OH-image in the same local position by using laser induced fluorescence (LIF). Figure 3 displays in the upper part the detected 2D signal which contains both signal contributions of the Rayleigh scattering and LIF as well. Using a recently developed detection and evaluation procedure, both contributions can be separated appropriately. In the middle part Figure 3 contains the 2D OH-field and in the lower part the temperature field being calculated from the Rayleigh signal. Details on the technique will be given elsewhere.

References:

BARAT, R.B., LONGWELL, J.P., SAROFIM, A.F., SMITH, S.P., BAR-ZIV, E. (1991): Appl. Opt. 30, 3003-3011

DIBBLE R. W., HOLLENBACH R. E. (1981): Proc. 18th Symposium (Int.) on Combustion, pp. 1489-1499. Pittsburgh: The Combustion Institute

DURST, F., ZARE, M. (1975): in Proc. LDA Symp., Copenhagen (DK) pp. 403-429,

DURST, F., MELLING, A., WITHELAW, J. H. (1976): Principles and Practice of Laser-Doppler Anemometry. London-New York: Academic Press

ECKBRETH, A. C. (1987): Laser Diagnostics for Combustion Temperature and Species. Cambridge (Mass.): Abacus Press

HAUMANN, J., LEIPERTZ, A. (1984): Opt. Lett. 9, 487-489

KAMPMANN, S., LEIPERTZ, A., DÖBBELING, K., HAUMANN, J., SATTELMAYER, T. (1993): Two-dimensional temperature measurements in a technical combustor using laser Rayleigh scattering, Appl. Opt. 32, in press

LEIPERTZ, A. (1989): in Instrumentation for Combustion and Flow in Engines, pp. 123-140, Dortrecht (NL): Kluwer Academic Pub.

LEIPERTZ, A. (1990): in Laser/Optoelectronics in Engineering, pp. 387 - 393, Heidelberg-New York: Springer Verlag

LEIPERTZ, A., KOWALEWSKI, G., KAMPMANN, S. (1992): in Proc. 7th Int. Symp.
on Temperature, pp. 685-690, New York: Am. Inst. Physics
LONG, M.B., CHANG, R. K., CHU, B.T. (1981): AIAA J. 19, 1151 - 1157
LONG, M. B., LEVIN, P. S., FOURGUETTE, D. G. (1985): Opt. Lett. 10, 267 -
269
MÜNCH, K.U., LEIPERTZ, A. (1992): SAE-Paper 922204

<u>Danksagung</u>

Teile der Arbeiten wurden durchgeführt mit Unterstützung der Deutschen
Forschungsgemeinschaft, im Rahmen des EG-Joule-IDEA-Programmes und
einer Zusammenarbeit mit Asea Brown Boveri (ABB-Baden/Schweiz).

Development and Test of a Mobile CARS System to Measure Temperature Fluctuatios in Large Industrial Combustion Systems

R. Bombach, B. Hemmerling, and W. Kreutner

Paul Scherrer Institute, Laboratory of Energy and Process Technology
CH - 5232 Villigen PSI, Switzerland

Coherent Anti-Stokes Raman Scattering (CARS) is a well established laser-based technique for non-intrusive, spatially resolved, determination of temperature and its fluctuations in flames (see e.g. [1-3]). The strong signal achievable with pulsed lasers and the coherent, laser-like nature of the signal beam render CARS a useful tool for measurements in large-scale flames. In this contribution, a mobile CARS system is described which has been developed at the Paul Scherrer Institute. Its ability to measure temperature is illustrated by some typical results obtained in an industrial burner.

The mobile CARS apparatus consists of a laser unit, a beam handling and delivery section, and a monochromator / detection unit. The laser unit includes the laser sources necessary for nitrogen CARS thermometry, their power supplies and the necessary beam alignment optics, in a solid frame of 2 x 0.8 x 1.35 m (L x W x H) outer dimensions. Temperature stabilisation and isolation against mechanical vibrations allow its employment even under harsh conditions. The system is built around a frequency-doubled Nd:YAG laser (Continuum NY81-20, running at a repetition rate of 20 Hz). A folded BOXCARS phase matching geometry is employed for signal generation in the flame. A part of the laser light is used to generate a non-resonant reference signal in a gas cell filled with butane at a pressure of 2 bar. The reference signal is focused into a glass fibre bundle and guided to the monochromator. The beam handling optics is installed in a frame made of aluminium beams. To allow measurements of temperature profiles across the flame, the lens (f = 200 mm) that focuses the laser beams into the burner can be translated on a motorised table by 200 mm. A second, identical device is used to move the recollimating lens. The resonant signal generated in the flame is focused into a second fibre bundle. The two bundles are merged in a "Y" junction and formed into a rectangular array of about 2 x

0.05 mm. The exit face of the combined fibre bundle is imaged onto the entrance slit of a SPEX model 1704 monochromator (1 m focal length, equipped with a 1800 l/mm grating). Resonant and non-resonant CARS signals are alternatively detected with a gated, intensified diode array camera (SI model S-IRY-1024). The translation of the measurement volume and the acquisition of reference and single-pulse resonant spectra is automated and runs under control of a PC program.

CARS measurements of temperature fluctuations were carried out in an industrial test rig equipped with a double cone burner developed by Asea Brown Boveri (ABB) [4], in a lean, turbulent, premixed natural gas / air flame burning at atmospheric pressure. The flame is stabilised near the burner outlet by utilising the sudden breakdown of the swirling flow upon exiting the burner cone [5]. The thermal load of the burner is about 120 kW under typical operating conditions. The burner air is electrically preheated to 650 K in order to simulate the effect of a compression stage. The air / fuel ratios λ employed are in the range between $\lambda = 1.9$ and 2.25.

In turbulent flames there are strong fluctuations of the CARS signal amplitude, which are caused both by the intrinsic temperature dependence of the CARS signal, and by beam steering effects. The decay time of the phosphor, used in the camera intensifier, sets a limit on the rate at which individual spectra can be read out sequentially, because a low intensity spectrum may be obscured by the residual lag from a preceding high intensity spectrum. However, in this burner the expected temperatures lie within the relatively narrow range of 650 K (preheated air) and 1690 K (adiabatic flame temperature) corresponding to an intensity change at the signal maximum of only about a factor of 30. Therefore, the maximum data acquiring rate of 20 Hz could be used for the single-shot spectra. A standard temperature measurement of one point within the radial profile consists of 1000 single-pulse spectra. A time-averaged, non-resonant reference spectrum is measured at the beginning and the end of the series and between two sets of single-pulse spectra. At the beginning of data collection a spectrum is recorded with the dye laser turned off. This spectrum contains mainly the electronic offset of the individual camera diodes and possibly some contribution of stray laser light. Before data storage, this background spectrum is subtracted from the recorded single-shot spectra. The data evaluation, consisting of two steps, is carried out off-line. First, all single-shot spectra are divided by a reference spectrum. In this way, the spectra were referenced to the averaged spectral profile of the dye laser, the individual relative sensitivity of the camera diodes, and the transmission characteristics of all optical components. Second, the temperature information is extracted from the flame spectra by comparison with a library of theoretical CARS spectra, pre-calculated for 25 K increments. The resulting temperatures are compiled in a normalised histogram with a 10 K bin width.

Single pulse temperature measurements by CARS exhibit the advantage of good spatial and temporal resolution. The temporal resolution, given by the duration of the laser pulse (approx. 10 ns), comfortably exceeds the demands encountered in the investigation of turbulent flames. The spatial resolution obtainable for the temperature measurement depends on the separation of the two pump beams in the BOXCARS configuration, and on the focal length of the lens which directs the three beams into the volume under investigation. Typically, a spatial resolution of 3 mm x 30µm is achievable with a 200 mm focal length of the focusing lens. Very often, this resolution is not good enough to resolve the flame front and the CARS amplitude is formed in a region containing both, hot and cold gases. Special temperature evaluation routines have to be employed in this case [6]. Single-pulse measurements in a stable, laminar laboratory flame yield for an isothermal medium a standard deviation of the histogram of 70 K at 2000 K. Furthermore, the accuracy of the CARS temperature measurement has been cross-checked in a graphite tube furnace by pyrometric temperature determinations. The discrepancy was smaller than 70 K at 2000 K.

Radial profiles of CARS temperatures were measured at various axial positions in the burner chamber. To illustrate the results, Figure 1 shows temperature probability density functions (pdf's) obtained at three locations along the burner axis. Near the burner outlet (normalised distance to the burner exit plane: $x/D = 0.1$), the pdf is rather broad and strongly skewed, with a

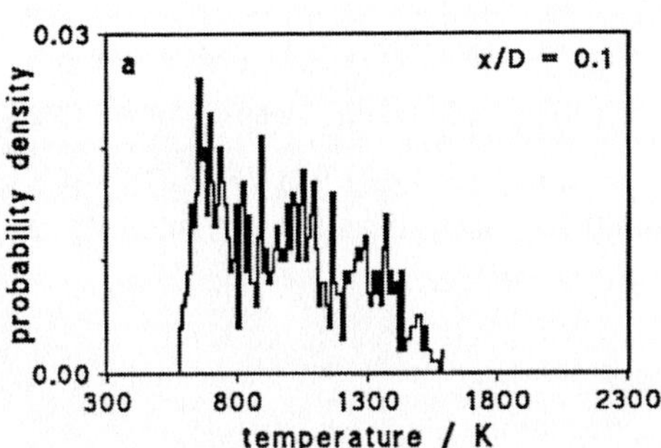

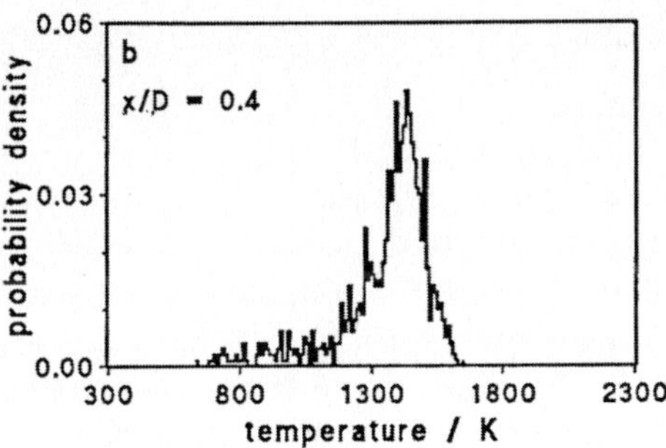

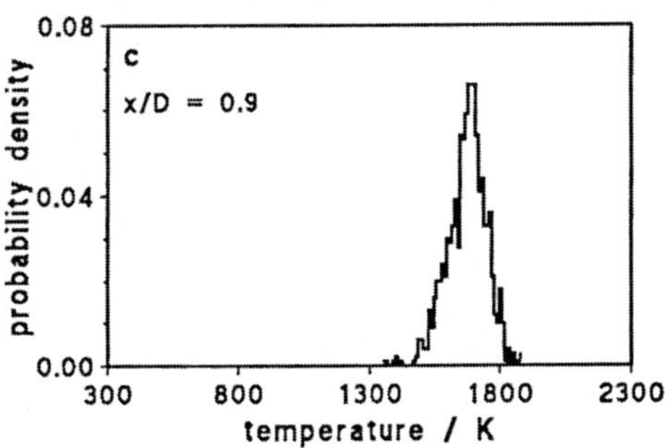

Figure 1. Single pulse CARS temperature probability density functions for a lean flame ($\lambda = 2.1$), measured on the burner axis at three axial positions.

maximum only slightly above the reactant temperature. Further downstream, at $x/D = 0.4$, the average temperature has increased substantially, the distribution narrows and shows a tail towards lower temperatures. At $x/D = 0.9$, the pdf is approximately Gaussian shaped with a mean temperature of $T_{av} = 1690$ K and a standard deviation of $\sigma = 70$ K, which coincides with the temperature measurement precision established in a laboratory flame. The three pdf's illustrate the progress of burn-out along the burner axis. At $x/D = 0.1$, only a fraction of the fuel has reacted, and the input temperature of the unburned reactants still occurs frequently. On the other hand, the absence of any measurement values below $T = 1350$ K in the pdf at $x/D = 0.9$ indicates the complete conversion of fuel. The narrow width and symmetric shape of this pdf show that the strongly turbulent flow produces a homogeneous exhaust gas region at this point on the burner axis.

Single pulse CARS measurements allow the determination of the temperature in a turbulent flame with a good temporal and spatial resolution. The repeated measurement at one point results in temperature pdf's. Probability density functions are adequate descriptions of a turbulent combustion process and are used for numeric modelling of flames. Work is in progress to extend the capabilities of the mobile CARS system in order to record simultaneously the temperature and the concentration of one of the major species in the flame.

Financial support by the Swiss Federal Office of Energy (BEW) is gratefully acknowledged. The field measurements were taken at the ABB Research Centre Baden-Daettwil, Switzerland; we thank the members of the CRBT1 group at ABB for the kind hospitality.

References

1) S.A.J. Druet, J.-P.E. Taran: Progr. Quantum Electronics Vol. 7, p.1, 1981
2) A.C. Eckbreth: "Laser Diagnostics for Combustion Temperature and Species", Abacus Press, 1988
3) D.A. Greenhalgh: "Quantitative CARS Spectroscopy", in: Advances in Non-Linear Spectroscopy, Vol. 15, p.193 (Eds.: R.J.H. Clark, R.E. Hester), Wiley, 1988
4) Th. Sattelmayer, M.P. Felchlin, J. Haumann, J. Hellat, D. Styner: "Second Generation Low-Emission Combustors for ABB Gas Turbines: Burner Development and Tests at Atmospheric Pressure", Paper 90-GT-162 presented at the ASME Gas Turbine and Aeroengine Congress and Exposition, June 11-14, 1990, Brussels, Belgium
5) J.J. Keller, W. Egli, R. Althus: ZAMP Vol. 39, p. 3, 1988
6) W. Meier, I. Plath, W. Stricker: Appl. Phys. B53, p.339, 1991

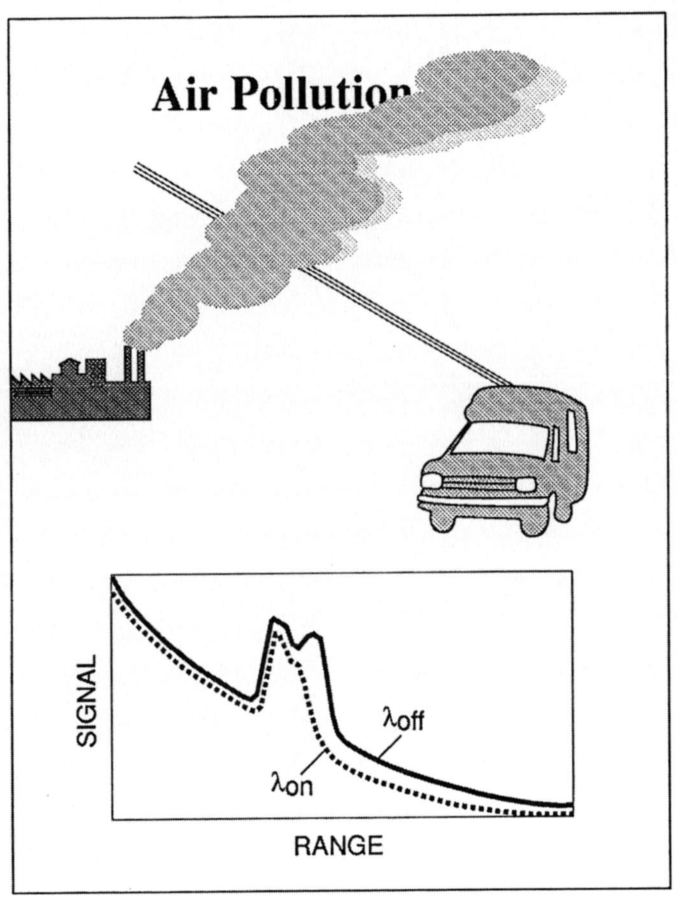

Air Pollution
SIGNAL
RANGE
λoff
λon

Environmental Analysis with the FT-IR System K300:
Examples and Results

T. Eisenmann[1], H. Mosebach[1], H. Bittner[1], M. Erhard[1], M. Resch[1], K. Schäfer[2], R. Haus[2]
[1]Kayser-Threde GmbH
Wolfratshauserstr. 44-48, D-8000 Munich 70
[2]Fraunhofer-Institute of Atmospheric Environmental
Research (IFU)
Kreuzeckbahnstr. 19, D-82467 Garmisch-Partenkirchen

1.   INTRODUCTION

Emission measurements of hot flue gases as well as ambient air quality monitoring have been performed with the FT-IR system K300 during different field tests in Berlin. These investigations have been conducted in the frame of a research project, funded by the German Ministry of Research and Technology in cooperation with the Fraunhofer-Institute of Atmospheric Environmental Research in Garmisch-Partenkirchen/Germany. Special emphasis has been put on the comparison of the results to conventional approved in-situ analyzers with respect to quality assurance. The FT-IR system and possible applications have been described in BITTNER et al., 1991 and MOSEBACH et al., 1991.

2.   DESCRIPTION OF THE APPLICATIONS

Results of the following two scenarios will be presented:

A) Flue gas emission measurements at power plants in Berlin. For this purpose a laboratory van has been equipped with a K300 and conventional approved in-situ analyzers. Besides the development of suitable evaluation procedures for the passive remote measurement of the hot smoke stack exhaust, an extended intercomparability study with continuous monitoring systems (CEM) installed in the power plants was to the fore. These comparative measurements were of great importance with respect to quality assurance to prove the equivalency of remote sensing results to well approved, widely accepted analytical systems. The principles of the analytical procedures as well as the results are described in detail in MOSEBACH et al. (1993, in prep.), HAUS et al. (1992) and EISENMANN et al. (1993).

B) Ambient air monitoring in the frame of the same research project at a clean air site and at a very busy intersection in the eastern part of Berlin. These measurements have been carried out with the mentioned laboratory van either. Included were comparative measurements to the in-situ analyzers in the van as well as to instruments of the Berlin Air Quality Monitoring Network (BLUME). Additionally, an intercomparison to canister sampling followed by gas chromatographic analysis in Garmisch-Partenkirchen has been carried out. These results and the principles of the concentration evaluation for transmission measurements used for ambient air control are depicted in MOSEBACH et al.(1993, in prep.) and EISENMANN et al. (1993).

3.   INVESTIGATIONS OF HOT FLUE GASES AT SELECTED POWER PLANTS IN BERLIN

Five different power plants have been investigated. Two were fired with pit coal, two with lignite coal and one with fuel oil. The mentioned laboratory van used for this purpose is described in detail in MOSEBACH et al. (1992). In the following chapters a few results of the intercomparison for the flue gas temperature, CO and NO will be depicted and briefly discussed.

3.1   Flue gas temperature

The temperature of the smoke stack exhaust has been determined using the CO_2-band around 2390 cm^{-1}. The determination has been described in MOSEBACH et al. (1993, in prep.).

Tab. 1: Comparison of temperature values, evaluted from IR-spectra (K300) and of an in-situ temperature sensor (IS). Values in °C. Fuel oil fired power plant, Nov. 3, 1992

K300	140	142	139	138	143
IS	150	149	147	150	147
DT	10	7	8	8	4

$$\delta T = -6.0 \text{ \%} \qquad \Delta T_{max} = -6.7 \text{ \%}$$

From this table it becomes obvious, that the temperatures derived from the in-situ sensor are always higher than those calculated from the CO_2-band. The maximum deviation was around 15 °C. Mainly two reasons are responsible for the observed differences. On the one hand the in-situ temperature is measured always at the bottom of the stack and the normal adiabatic cooling of the upstreaming flue gas accounts for a decrease of 5 °C on the average. On the other hand an immediate cooling takes place at the moment the flue gas is leaving the stack due to the expansion and a rapid mixing with the atmosphere.

The mean deviation between the two different temperatures, calculated from all stack measurements, lies around 5.5 % and represents a quite good agreement. After subtraction of the adiabatic cooling , the agreement for a lot of measurements was better than 1 %.

3.2   Carbon Monoxide

In tab. 2 an intercomparison is depicted between a continuous emission analyzer and the K300.

Tab. 2: Comparison between the K300 and an in-situ analyser (IS) for CO. Concentrations in mg/m^3, pit coal, Nov. 10, 1992

K300	193	157	156	169	165
IS	200	160	180	185	195
DQ	7	3	24	16	30

$$\delta Q = -10.7 \text{ \%} \qquad \Delta Q_{max} = -15.4 \text{ \%}$$

For a correct evaluation of the flue gas concentration in the plume the knowledge of the atmospheric concentrations in the foreground between the plume and the instrument is of great importance, especially if these concentrations are fluctuating as in the case of CO. If these concentration values are assumed or determined too low, the spectral simulation yields too high stack concentrations, because the absorption of the radiation originating from the plume constituents by the corresponding compound in the foreground will be underestimated. To achieve as precise foreground concentrations as possible the in-situ analyzers in the van have been used.

3.3   Nitrogen Monoxide

The determination of NO is rather difficult due to the strong influence of the atmospheric water vapor in the relevant spectral region at 1900 cm^{-1}. Table 3 and 4 show an intercomparison for relatively high and low NO-concentrations (near the detection limit of 20 mg/m^3 valid for these measurement conditions).

Tab. 3: Comparison of K300 and in-situ (IS) NO-concentrations in mg/m³, pit coal, Sep. 30, 1992

K300	566	611	541	523	622
IS	527	525	507	519	513
DQ	39	86	34	4	109

$$\delta Q = +9.9 \ \% \qquad \Delta Qmax = +17.5. \ \%$$

Tab. 4: Comparison of K300 and in-situ (IS) NO-concentrations in mg/m³, fuel oil, Aug. 6, 1992

K300	20	24	26	20	27
IS	59	59	59	59	59
DQ	39	35	33	39	32

$$\delta Q = -61 \ \% \qquad \Delta Qmax = -66.1 \ \%$$

3.4   Assessment of the results

Due to numerous factors influencing the concentration evaluation (MOSEBACH et al., 1993, in Vorb.) und the elusive influence of the wind, an expected agreement between the CEM's and the FTIR better than 50 % was regarded to be satisfactory. In reality the agreement is much better, if averaged over all evaluated results, even for difficult gases like NO. On the average the following differences between the in-situ data and the FTIR results for the main important constituents could be achieved:

$$CO : - \ 7,8 \ \%$$
$$NO : - \ 11.5 \ \%$$
$$CO_2: + \ 1.2 \ \%$$

The agreement of the flue gas temperature is in many cases much better, as mentioned above.

4.   AMBIENT AIR MONITORING AT DIFFERENT POLLUTED SITES IN BERLIN AND GARMISCH-PARTENKIRCHEN/GERMANY

Another major objective of the above mentioned research project was to monitor the ambient air concentration levels of CO, CO_2, NO_x, N_2O, CH_4 und Formaldehyd (HCHO) at a clean air site and a very busy intersection in the eastern part of Berlin. During these investigations the mentioned intercomparability study to analyzers of the Berlin Air Quality Monitoring Network (BLUME) and the point sensors in the laboratory van has been carried out for CO and NO.

These studies have been supplemented by comparative measurements of the FTIR and gas chromatography of canister samples for CO, CO_2, N_2O and CH_4.

For all these measurements a so-called bistatic transmission configuration with an artificial infrared source (glowbar) facing the instrument has been used. The pathlengths were up to several hundred meters.

For CO_2, N_2O and CH_4 the evaluated concentrations were in quite good agreement with the atmospheric background concentration levels, reported in the literature (CO_2: 320-330 ppm for clean air sites, 340-380 ppm for urban areas. N_2O: around 320 ppb. CH_4: 1.6 ppm, in urban areas 10-20 % higher). The degree of agreement can be regarded as a general plausibility check for the measurements itself.

4.1   Comparison with conventional gas analyzers

The two in-situ CO-analyzers used the principle of nondispersive infrared photometry (NDIR) and the two NO-instruments chemiluminescence. All these instruments were officially approved for ambient air control. The different in-situ systems were located at the opposite ends of the measurement path, which means, that the infrared source stood next to the BLUME container, the other in-situ instruments were located next to the FTIR in the van. The measurement time was synchronized in order to achieve corresponding half hour mean values.

From table 5 below it becomes apparent, that the results of the FTIR tend to be higher in most cases. A possible reason for this might be, that the air monitored by the NDIR analyzers was sampled at a height of 3 or 3.8 m above the street level whereas the measurement path ran in a height between 2.5 m (FTIR) and 1.3 m (glowbar). Hence both air samples were subject to a dilution with regard to the air volume measured along the FTIR measurement path. Nevertheless this is not valid for all results and additionally one has to consider, that the both in-situ systems analysed different air masses. In such cases a measurement technique achieving path-averaged results has distinct advantages by smoothing small-scale concentration differences and fluctuations due to local sources.

Tab. 5: Comparison between the two NDIR CO-analyzer and the K300, Aug. 25, 1992, near a busy intersection in the eastern part of Berlin. Half hour mean values in ppm.

Time-period	K300	CO-analyzer van	CO-analyzer BLUME
13:00-13:30	1.2	0.3	0.6
13:30-14:00	1.3	0.5	0.8
14:00-14:30	1.1	0.4	1.0
14:30-15:00	1.0	0.5	1.4
15:00-15:30	1.7	1.1	2.2
15:30-16:00	1.8	1.4	2.6

Tab. 6: Comparison between two chemiluminescence NO-analyzer and the K300, Aug. 26, 1992 near a busy intersection. Half hou mean values in ppb.

Time period	K300	NO-analyzer van	NO-analyzer BLUME
10:00-10:30	63	76	47
11:30-12:00	81	74	45
12:00-12:30	72	51	43
12:30-13:00	66	31	40
13:00-13:30	26	35	40
16:30-17:00	34	21	17
17:00-17:30	51	40	9

As depicted in the above table the FTIR values tend to be higher with respect to the in-situ results as well.

Probably the same reason could be responsible for this phenomenon as for CO, but here the difference is more distinct. Generally the interpretation of the NO results is aggravated due to photochemical reactions and the formation of ozone.

For a better assessment of the comparability of such different measurement approaches a similar study has to be conducted at a clean air site. There homogeneous air masses with equally dsitributed concentrations can be expected. Within the scope of the above described investigations there was no possibility for this kind of comparison, but it has been realised with the below depicted canister measurements to a large degree, because the selected site in Garmisch-Partenkirchen showed nearly background concentration levels.

4.2 Comparison of the FTIR to canister sampling and GC analysis

Along a measurement path with a length of 250 m three canister samplers have been placed at distances of 25, 125 and 225 m away from the laboratory van. 8 sample sets, each consisting of three averaged sub-samples have been analysed in the laboratory with a GC-system for CO_2, CO, N_2O and CH_4. These measurements have been performed by the Fraunhofer Institute of Atmospheric Environmental Research in Garmisch-Partenkirchen. The agreement between the FTIR and the laboratory results is fairly good for all the gaseous species despite relatively large differences between the single canisters in some cases. As an example the comparison for CO is depicted in table 7.

Tab. 7: Comparison of K300 and canister measurements for CO. All values are given in ppb.

K300	Canister			Average Kanister 1-3	DQ
	1	2	3		
190	212	174	204	197	7
180	194	177	185	185	5
192	189	187	189	188	3
169	152	194	154	167	2
158	166	166	164	165	7
163	166	169	205?	180	17
153	157	153	142	151	2
180	176	175	181	177	3
	Average				
173	176	174	178	176	3

$$\delta Q^E_{K300} = +\ 5\ ppb \qquad \delta Q^E C_{an} = +\ 4\ ppb \qquad \delta Q^{AV} = -1.7\ \%$$

4.3 Limitations of the long-path monitoring for the measured gases

Most of the gases, investigated within the described research project have detection limits much below their ambient air concentration levels. Only the determination of NO is difficult due to the severe spectral interference of the atmospheric water vapor. Additionally, the background value is far below those of the other mentioned gases and therefore it could be detected at polluted sites only. The detection limits can be improved by applying longer measurement paths due to the BEER-LAMBERT law. Nevertheless this is actually valid for gases not influenced by atmospheric water vapor and CO_2 to a higher degree. This leads to certain limitations for e.g. NO and SO_2.

In general, the application of a high spectral resolution has distinct advantages for gases exhibiting line spectra in the infrared spectral region despite the associated deterioration of the signal-to-background ratio. Again

NO may serve as a good example. Comparative measurements with the K300 showed, that almost no chance exists to detect NO by applying a spectral resolution of 1 cm^{-1}, even at very high ambient concentration levels (e.g. at heavily polluted sites).

5. REFERENCES

BITTNER, H., ERHARD, M., NEUREITHER, I., MOSEBACH, H. & RIPPEL, H. (1991): The K300 Fourier Transform Spectrometer, Environmental Applications of the Double Pendulum Interferometer - Proceedings of the 8th International Conference on Fourier Transform Spectroscopy, SPIE-1575, S. 186-188, Society of Photo-Optical Instrumentation Engineers, Washington.

EISENMANN, T., MOSEBACH, H., BITTNER, H., HAUS, R. & SCHÄFER, K. (1993): Selected Applications of Remote Sensing Measurements with the Double Pendulum Interferometer in Germany with Special Consideration of Quality Assurance Aspects - 31 pp., oral presentation at the 6th annual meeting and exhibition of the Air & Waste Management Association, June 14-18, Denver.

HAUS, R., SCHÄFER, K., WEHNER, D., MOSEBACH, H. & BITTNER, H. (1992): Remote Sensing of Air Pollution by Mobile Fourier- Transform Spectroscopy: Modeling and first Results of Measurements - Proceedings of the 1992 International Speciality Conference on Optical Remote Sensing Applications to Environmental and Industrial Safety Problems, SP-81, S. 67-75, Air & Waste Management Association, Pittsburgh.

MOSEBACH, H., BITTNER, H. & RIPPEL, H. (1991): Einsatzmöglichkeiten des Doppelpendelinterferometers DPI zur Emissions- und Immissionsmessung von Luftschadstoffen - Laser in emote Sensing (WERNER, C., KLEIN, V. & WEBER, K., eds.), Proceedings of the 10th International Congress, Laser '91, pp. 221-232, Springer.

MOSEBACH, H., EISENMANN, T., SCHULZ-SPAHR, Y., NEUREITHER, I., BITTNER, H., RIPPEL, H., SCHÄFER, K., WEHNER D. & HAUS, R. (1992): Remote Sensing of Smoke Stack Emissions using a Mobile Environmental Laboratory - Industrial, Municipial and Medical Waste Incineration Diagnostics and Control, SPIE-1717, Society of Photo-Optical Instrumentation Engineers, Washington

MOSEBACH, H., SCHÄFER, K., HAUS, R., EISENMANN, T., WEHNER, D., RIPPEL, H. & KLEIN, V. (1993): MEISTER - Development and Testing of a Mobile Measurement System for the former German Democratic Republic for Monitoring Ambient Air and Emissions of Industrial Plants - 149 pp., final report BMFT 01 VQ 9025, Kayser-Threde GmbH, Munich

Verfahren zur passiven und aktiven Fernsondierung von Spurengasen auf der Basis der Fourier-Transform-Infrarot-Spektroskopie

K. Schäfer*, R. Haus+, W. Bautzer+, H. Mosebach#, H. Bittner#, T. Eisenmann#

*Fraunhofer-Institut für Atmosphärische Umweltforschung (IFU),
Kreuzeckbahnstr. 19, 82453 Garmisch-Partenkirchen
+Rudower Chaussee 5, 12489 Berlin
#Kayser-Threde GmbH, Wolfratshauser Str. 44-48, 81379 München

Zusammenfassung

Das Meßfahrzeug mit dem Doppelpendel-Interferometer K 300 der Fa. Kayser-Threde wurde erfolgreich getestet und zur Analyse von Abgasen sowie der Umgebungsluft operationell eingesetzt. Die gemessenen Infrarot-Spektren von natürlichen oder künstlichen Strahlungsquellen werden mit PC-Programmen auf der Basis der Strahlungstransporttheorie ausgewertet. Damit können aus der Emission oder Absorption der Strahlung die Konzentrationen von CO, N_2O, CH_4, CO_2, H_2O, NO, NO_2, NH_3, HNO_3, SO_2, O_3, HCl und $HCHO$ in dem entsprechenden Medium bestimmt werden. Es werden Beispiele zum Auswerteverfahren für einen Kraftwerksschornsteinen und eine Luftmeßstation dargestellt.

Einleitung

Die Nutzung der Fourier-Transform-Infrarot-Spektrometrie (FTIR) als passives und aktives Verfahren zur Bestimmung der Konzentration von Spurenkomponenten in der Atmosphäre erfolgt seit den 50-er Jahren. Die technischen Entwicklungen sind jetzt so weit fortgeschritten, daß dieses Meßverfahren herkömmliche Verfahren ergänzen und in naher Zukunft ablösen kann. Die Gründe dafür sind folgende:
- Bestimmung eines vollständigen Spektrums aus einer einzelnen Strahlungsmessung, so daß eine Multikomponentenanalyse erfolgt,
- Erfassung aller klimarelevanten Gase,
- Bestimmung von Luftschadstoffen und insbesondere organischen

Komponenten (VOC) in Emissionskonzentrationsbereichen (zur Zeit ca. 150 in Spektrenkatalogen aufgelistet),

- Durchführung passiver Messungen von warmen Gaswolken und

- Bestimmung der mittleren Konzentration über einige 100 m mittels Absorptionsmessungen mit einer Strahlungsquelle, so daß lokale Inhomogenitäten und Luftturbulenzen keine Rolle spielen sowie eine Anpassung an Gitterweiten von Ausbreitungsmodellen möglich ist.

Meßverfahren

Die Verfahrensentwicklung hatte ein operationell einsetzbares Meßsystem zum Ziel und war Gegenstand eines vom BMFT geförderten Projektes. Es wurde ein Meßfahrzeug entwickelt, in dem als Strahlungsmeßgerät das Doppelpendel-Interferometer K 300 der Firma Kayser-Threde eingesetzt wird /MOSEBACH, EISENMANN et al., 1992/. Mittels Teleskop empfängt das Meßsystem sowohl die Eigenstrahlung von warmen Gaswolken als auch das Licht einer Strahlungsquelle, was mit keinem anderen kommerziellen transportablen FTIR-Gerät möglich ist. Die spektrale Auflösung des Meßsystems ist variabel bis 0,06 cm^{-1} /BITTNER, ERHARD et al., 1991/.

Die Strahlung einer Gaswolke wird wesentlich von den Strahlungseigenschaften des Vorder- aber auch des Hintergrundes beeinflußt. Die erstellten Auswertealgorithmen berücksichtigen diese Verhältnisse sowie beliebige Temperatur- und Druckbedingungen /HAUS, GOERING et al., 1991/, so daß die Vorteile des K 300 voll genutzt werden. Das Auswerteprogramm baut auf der Modellierung des Strahlungstransportes mit einem früher entwickelten Linie-für-Linie Programm auf HITRAN-Spektrallinien-Basis /ROTHMAN, GAMACHE et al., 1987/ auf, womit synthetische Spektren berechnet werden. Damit können die komplizierten Schichtungsverhältnisse erfaßt werden, was bei Nutzung von Spektrenkatalogen nicht möglich ist /SCHÄFER, HAUS et al., 1992/. Von der Auswerte-Software wird dann das Eingangsmodell über die Zusammensetzung des Untersuchungsobjektes durch Vergleich der berechneten mit den gemessenen Spektren (least square fit) solange modifiziert, bis eine Übereinstimmung bis auf Abweichungen in der Größenordnung des Meßfehlers erreicht ist. Diese PC-Programme wurden bei der

Erprobung des Meßsystems eingesetzt und optimiert. Für die verschiedenen Meßaufgaben mußten aufwendige Untersuchungen über die Querempfindlichkeiten der zu bestimmenden Komponenten (siehe Tabelle) und die Algorithmen der Konzentrationsbestimmung durchgeführt werden /HAUS, SCHÄFER et al., 1992/.

Tabelle: Optimale Spektralregionen (in cm^{-1}) für die zu bestimmenden Komponenten und Querempfindlichkeiten mit anderen Stoffen für Emissions- (E) und Absorptionsmessungen (A)

Komponente	Querempfindlichkeiten	Spektralregion	Meßkonfiguration
H_2O	-	1990-2010	E
	-	1965-1985	A
	-	2655-2675	A
CO_2	H_2O, CO	2045-2065	E, A
	-	2390-2400	E
N_2O	H_2O, CO, (CO_2)	2180-2200	E
	H_2O, CO	2180-2200	A
CO	H_2O, N_2O	2160-2180	E
	H_2O, CO_2	2090-2110	A
CH_4	H_2O	3030-3050	E, A
	H_2O, N_2O	1295-1315	E, A
NO	H_2O, CO_2	1890-1910	E, A
SO_2	H_2O	2500-2520	E, A
	H_2O	1150-1170	E, A
NO_2	H_2O	2910-2930	E, A
	H_2O	1590-1610	E, A
NH_3	H_2O, CO_2	920- 940	E, A
HCl	H_2O	2765-2785	E, A
HCHO	H_2O	2770-2790	E, A

Erprobung des Meßverfahrens

Die Erprobung des Meßsystems erfolgte mit Testmessungen an Kraftwerksschornsteinen und trassengemittelten Absorptionsmessungen in Bodennähe. In den Abbildungen 1 und 2 sind für Beispiele beider Meßkonfigurationen die Ergebnisse der Anpassung der simulierten an die gemessenen Spektren sowie die in den ausgewählten Spektralbereichen noch vorhandenen Querempfindlichkeiten dargestellt. Es ist auch der Fall mit der Vernachlässigung anderer Komponenten in der Simulation gezeigt, was z. B. beim NO zu Fehlern von 100 % führt.

Für die Analyse von Schornsteinabgasen wird das Teleskop direkt über den Schornsteinkopf gerichtet und empfängt die Strahlung, die von der Abgaswolke ausgesandt wird. Die Dicke der Abgasfahne wird durch Schwenken des Eingangsspiegels vom einen zum anderen Rand der Fahne ermittelt. Die Entfernung zum Schornsteinkopf wird mit einem Laserentfernungsmesser bestimmt. Die

Schadstoffbelastung im Vordergrund wird aus einer Absorptionsmessung mit einer Infrarot-Strahlungsquelle errechnet. Mittels FTIR ist es allerdings nicht möglich, die Strömungsgeschwindigkeit und die Sauerstoffkonzentration zu messen.

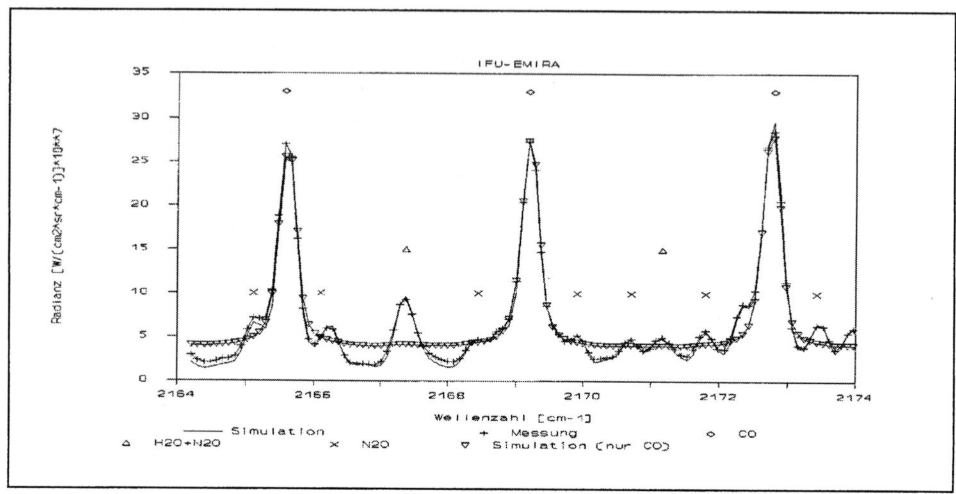

Abb. 1: Least square fit für ein CO-Spektrum aus einer Emissionsmessung an einem Steinkohle-Kraftwerk

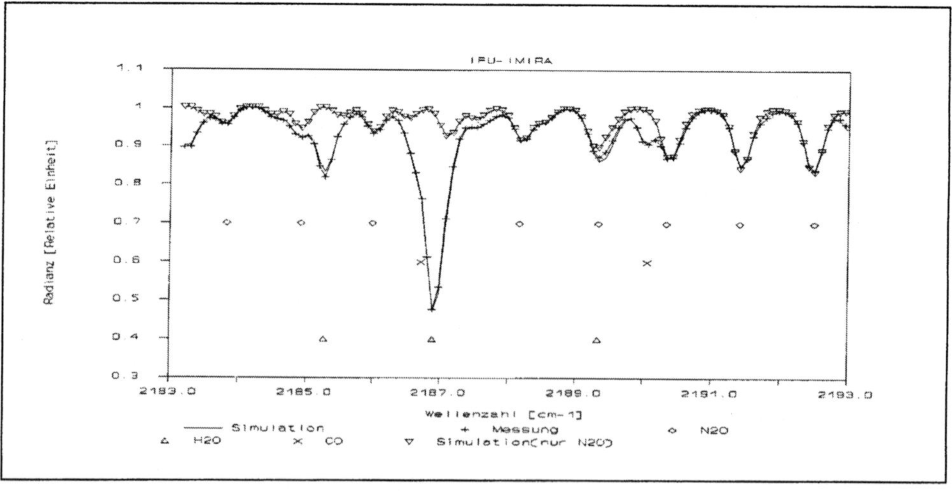

Abb. 2: Least square fit für ein N_2O-Spektrum aus einer Absorptionsmessung

Es wurden die Abgase von Kraftwerken untersucht, die mit Braunkohle, Steinkohle und Heizöl betrieben werden. Diese

Kraftwerke verfügen über keine Entstickungsanlagen und nur zum Teil über eine Entschwefelung. Es wurden die Konzentrationen von CO_2, CO (siehe auch Abb. 1), NO, N_2O, HCl und H_2O sowie die Temperatur des Abgases bestimmt. Bei Abgasen, die eine Rauchgasentschwefelung passiert hatten, konnte kein SO_2 mehr nachgewiesen werden. Die emittierte Gesamtkonzentration von Kohlenwasserstoffen lag bei allen Kraftwerken unter der Nachweisgrenze von 1 ppm.

Als weitere Emissionsquelle wurde der Straßenverkehr untersucht. Dazu wurde die Fourier-Transform-Spektrometrie aktiv, d. h. unter Nutzung einer Strahlungsquelle für eine Absorptionsmessung eingesetzt. Der Vergleich mit Meßcontainern des BLUME-Meßnetzes zeigte, daß die Punktmessung wesentlich stärker meteorologischen Einflüssen und Inhomogenitäten der Emission ausgesetzt ist als die Trassenmessung.

An der Intensivmeßstation des SANA-Programms in Melpitz erfolgten während charakteristischer Wetterlagen 1- bis 2-tägige Meßkampagnen. In Abb. 2 ist das Ergebnis der Anpassung von synthetischem und gemessenem Spektrum dargestellt. Die Messungen ergaben, daß CO im Gegensatz zu CO_2, N_2O und CH_4 durch ausgeprägte Schwankungen im Laufe des Tages charakterisiert ist. Dieses Verhalten zeigt auch die jahreszeitliche Variation. CO ist im Winter um den Faktor 2 höher als im Sommer und CO_2 sowie N_2O sind durch einen leicht ansteigenden Trend charakterisiert.

Leistungsfähigkeit des Meßverfahrens

Die FTIR-Ergebnisse wurden mit in-situ Messungen im Rauchgaskanal bzw. der Analyse von Luftproben evaluiert. Das Meßsystem verfügt über folgende Merkmale im operationellen Einsatz /MOSEBACH, SCHÄFER et al., 1993/:

- es ist mobil, und für den Aufbau aller Systemkomponenten zur Durchführung der Messung wird rund eine Stunde benötigt,
- eine Messung dauert wenige Sekunden,
- die Datenauswertung erfolgt on-line und erfordert wenige Minuten, so daß Messung und Auswertung weniger als 5 Minuten benötigen,
- Emissionsmessungen können mit einer Genauigkeit von besser als 20 % für folgende Gase mit den entsprechenden Nachweisgrenzen bestimmt werden: CO 3 ppm, NO 10 ppm, N_2O 10 ppm, HCl 1 ppm, NO_2

80 ppm, SO_2 400 ppm, CO_2 0,1 % und H_2O 2 %; zusätzlich wird die Temperatur der Gaswolke ermittelt,

- die Nachweisgrenzen des Verfahrens ermöglichen die Bestimmung der Konzentrationen von CO, CO_2, CH_4, N_2O und H_2O unter beliebigen Luftbelastungsbedingungen bis hin zu Reinluftgebieten mit einer Genauigkeit besser als 7,5 %,

- bei hoher Belastung können zusätzlich NO, HCHO, NH_3 und HNO_3 mit einer Nachweisgrenze von jeweils 10 ppb sowie NO_2, SO_2, HCl und O_3 aus der gleichen Messung bestimmt werden und

- die Messung ist nicht möglich bei starkem Regen oder Schneefall.

Literatur

Bittner, H., M. Erhard et al.: "The K 300 Fourier Transform Spectrometer", SPIE, 1575 (1991).

Haus, R., H. Goering: "Atmosphärenphysikalische Grundlagen der infrarotspektroskopischen Luftanalyse", Report des Heinrich-Hertz-Instituts für Atmosphärenforschung und Geomagnetismus der Akademie der Wissenschaften, Berlin 1991.

Haus, R., K. Schäfer et al.: "Remote Sensing of Air Pollution by Mobile Fourier Transform Spectroscopy: Modelling and First Results of Measurements", Proceedings of the AWMA International Specialty Conference on Optical Remote Sensing and Application to Environmental and Industrial Safety Problems, Houston, Air and Waste Management Association, Pittsburgh, PA, SP-81, 67-75 (1992).

Mosebach, H., T. Eisenmann et al.: "Remote sensing of Smoke Stack Emissions Using a Mobile Environmental Laboratory", SPIE, 1717 (1992).

Mosebach, H., K. Schäfer et al.: "MEISTER - Entwicklung und Erprobung eines mobilen Meßsystems zur Konzentrationsbestimmung geführter Emissionen sowie Immissionen von Massenschadstoffen in der ehemaligen DDR", Abschlußbericht des Projektes 01 VQ 9025, München 1993.

Rothman, L. S., R.R. Gamache et al.: "The HITRAN database, 1986 edition", Appl. Opt., 26, 4058 (1987).

Schäfer, K., R. Haus et al.: "Modelling of Radiative Transfer for FTIR Remote Sensing of Smoke Stack Emissions", Proceedings of the 9th World Clean Air Congress, Montreal, Air and Waste Management Association, Pittsburgh, PA, 1992, Paper IU-9B.05.

Single-Longitudinal-Mode (SLM) Generation at Power Diode Lasers by Using a Diffracto-Optical External Cavity

Hartmut G. Hänsel   ReFIT e. V., Selierstr.6, D - 07745 Jena
Rolf Heilmann   DLR, Institut f. Optoelektronik, D - 82230 Wessling
H.-J. Dobschal, Klaus Rudolf Carl Zeiss JENA GmbH, Selierstr. 6, D - 07745 Jena

Motivation

Environmental research at the atmosphere contains a lot of methods investigating the chemical nature of air pollutions and their distribution in the air. Laser spectroscopic methods and Lidar among these are able to solve some of that tasks. Laser analytical investigations make necessary tunable laser sources. For mobile operations at satellites or aeroplanes a lidarsystem has to be small, light, compact, robust and reliable. The improvement of such systems could be substantial by using laser diodes. In comparison with conventional types of lasers (CO_2- or solid state lasers) laser diodes have the advantage of rather simple tunability and modulability.

An efficient Doppler-lidar-system requires more than 1 Watt output of the laser diode by realizing SLM-(single-logitudinal-mode-) regime. For measuring wind-velocities for instance with accuracies of about 1 $m\,s^{-1}$ by using AlGaAs-laser diodes ($\lambda \approx 800$ nm) spectral linewidths $\Delta\lambda$ of less than 3 MHz are aimed at. By using conventional resonator geometries such values are obtained only for output-powers at the milliwatt-region. To improve the output-power in realizing SLM-regime currently there are two basic concepts: MOPAs (master oscillator power amplifier) and external cavities. In this paper we concern with external cavities.

External-cavity-characteristics

External cavities are known to narrow /1/ the basic spectral linewidth $\Delta\lambda_0$ of laser diode radiation to $\Delta\lambda$ values aimed at. Different geometries of external cavities are used. The basic /2/ and most convenient /3/ one comprises a plan grating connected to collimating optics and some other correcting optics if needed. How to see at the formular below, the linewidth narrowing is caused by two facts: a long external cavity L_{est} and a strong external coupling coefficient k_{ext} together realizing a strong feedback-coefficient C. The coupling k_{ext} is mainly determined by the ratio of reflectivities of the external cavity to that of the laser-output-facet. This means, that on the one hand a low antireflection-coating is necessary, on the other hand a high external reflectivity. Because the latter one includes the power capturing capability of cavity-optics by imaging properties, especially for power laser diodes a feedback from all of the laser stripes to all of them has to be realized in an efficient way.

$$LIF = \frac{\Delta v}{\Delta v_0} = \frac{1}{(1 + C)^2}$$

$$C = \frac{\tau_{ext}}{\tau_L} \, k_{ext} \, \sqrt{1 + \alpha^2}$$

$$k_{ext} = \frac{r_{2\,ext}}{r_{2\,s}} \left(1 - |r_2|^2\right)$$

LIF	=	Linewidth-Improvement-Factor
C	=	feedback-coefficient
α	=	chirp-parameter
Δv	=	laser-linewidth to be aimed at with cavity
Δv_0	=	basic-laser-linewidth
τ_{ext}	=	round-trip-delay (forth and back) through the external cavity of length L_{ext}
τ_L	=	round-trip-delay of the solitary laser diode of length L
k_{ext}	=	external-coupling-coefficient
r_{2ext}	=	reflection-coefficient of the external cavity
r_{2s}	=	reflection-coefficient of the solitary laser diode

In conventional manner one has to make a hard effort, because the better the imaging is got the greater the losses in the multitude of optical correcting elements needed for it.

We have overgone these difficulties by utilizing a diffractro-optical /4/ external cavity, consisting of only one diffractro-optical element (DOE), which is holographically made, containing all the optical functions as reflection, collimation, wavelength dispersion and especially imaging correction at one surface. In a first attempt we have applied this principle to given laser diodes as described below.

Experimental arrangement

Fig. 1 shows a scheme of the experimental arrangement. Aim of the investigation is the principal verification of SLM-generation at power laser diodes connected to a DOE, realizing an external cavity. This should be done at the first step without optimized elements, only with such ones available at that time. The arrangement is a very simple one:

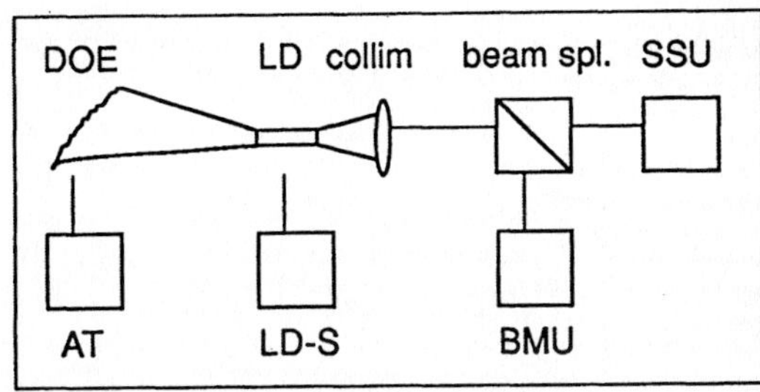

Fig. 1: Scheme of experimental arrangement

The DOE is adjusted in relation to the laser diode (LD) by a "10-degrees of freedom" adjustment-tower (AT). The laser diode in the laser head is supplied S with laser current, but without thermical stabilization. Laser radiation is fed by a collimating optics to a beam splitter, which gives on part to a beam monitoring unit (BMU) and another part to a survey-spectral unit (SSU). DOE is an element, which is regionally optimized by Carl Zeiss Jena GmbH for utilizing laser diodes VQ150 (of WF-Berlin) at 842 nm, totally unfitted to the 812 nm-emitting laser diodes, how to see at Fig. 4 later. The AT of OWIS realizes 6 degrees of freedom roughly (3 translation, 3 rotation) and 4 ones fine (3 translation, 1 rotation) piezoelectrically moved. As laser diode under test serves alternately a SDL-2460-C of SDL and a Siemens-LD SFH 48E1. Our laser diodes are arranged in such a manner in the external cavity, that the originally output (with mainly undisturbed wavefronts) is directed towards the DOE and the other side (originally directed to monitor diode) serves as output. Therefore the Siemens LD had to be mechanically prepared. Regarding to the coating the LDs are used unchanged,without preparations. This means the r_{2s}-value is about 32 % (instead of less than 1 % optimized) and the "new-output"-facet was estimated (according a method of Sigg /5/) to have a reflectivity of more than 90 % (SDL: 98.9 %; Siemens: 93.5 %). Spectral emission is observed to lie in the region of greater 812 nm and to have a linewidth of about 1 nm. Laser supplier is a SDL 800 M of SDL. The BMU consists of a TV-camera and monitor, the SSU consists of telescope, beam splitter, attenuator, plane grating, CCD line camera and oscilloscope, where the spectral behaviour of the radiation can be observed with tenth of nanometers resolution.

Results

During experiments the SLM-generation at power laser diodes by using a diffracto-optical external cavity is verified, despite the DOE is spectral unfitted to LD. Results displayed for SFH 48E1 laser diode.

According Fig. 2 a for laser current I_{LD} = 551 mA ($I_{th} \approx$ 280 mA) the gain is observed at 819.5 nm and (after switching "on" the external cavity) the generated SLM at 821 nm. The single-mode power exceeds the multi-mode one by a factor of 1.3 till 1.4. Fig. 2b shows the SLM generated SLM at 823 nm for 701 mA and the gain of unswitched cavity at 821 nm. The linewidth is in both cases about 0.2 nm. Single-mode power again exceeds the multi-mode one. The SLM-radation is tunable too. At Fig. 3 the radiating laser stripes are seen. Depending on the state of adjustment each of them is operating more or less. In the observed cases of SLM-regime according Fig. 2 all stripes are almost equally in operation.

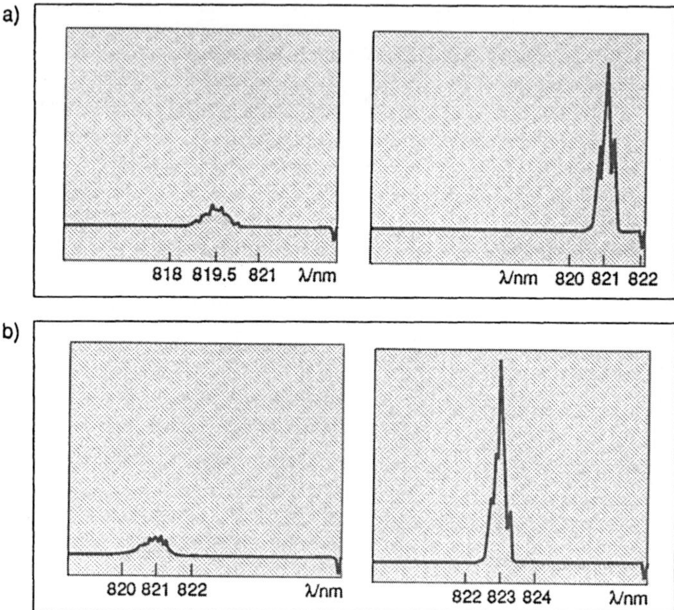

Fig. 2: a) Spectral behaviour of laserdiode-radiation at laser current 551 mA without and with diffracto-optical external cavity at 819.5 nm and 821 nm (power in arbitrary units)

b) Spectral behaviour of laserdiode-radiation at laser current 701 mA without and with diffracto-optical external cavity at 821 nm and 823 nm (power in arbitrary units)

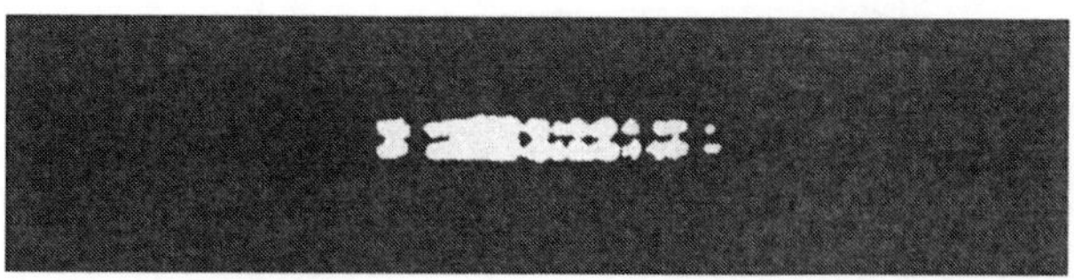

Fig. 3: Radiating laser stripes (together with the mirrored image)

Further prospects

After having fitted the DOE according the spectral and spatial requirements of the given LD the feedback-imaging can be improved, by factors, how to see at Fig. 4. A 160 points spot-diagramm is displayed in
Fig. 4a showing the image of one central laser-stripe point diffracted at
160 points, distributed over the whole surface of the DOE realizing
that image on the laser facet. The resulting difference between fitted
(optimized for 812 nm) and unfitted (optimized for 842 nm) DOE is
seen drastically. Fig. 4b similarly displays five points along the 100-
mikron emitting area (for instance of a 12-stripe power laser diode)
diffracted versus DOE to itself. Stabilizing the thermal conditions of
the laser diode and mechanical stabilization of the whole arrangement are further tasks as well as going over to eye-safe (lidar-) wavelengths.

a)

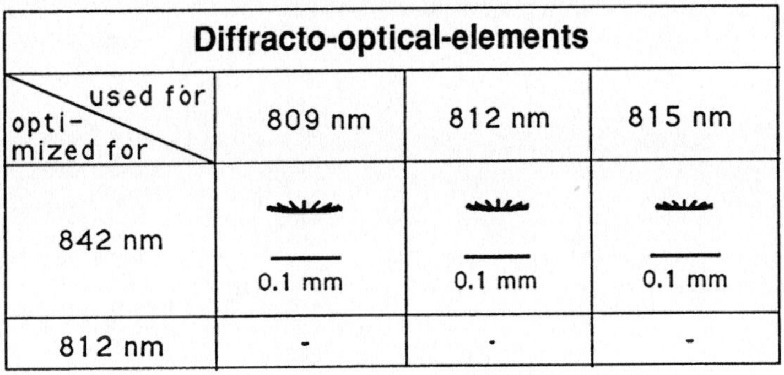

Diffracto-optical-elements			
used for / opti-mized for	809 nm	812 nm	815 nm
842 nm	0.1 mm	0.1 mm	0.1 mm
812 nm	-	.	-

b)

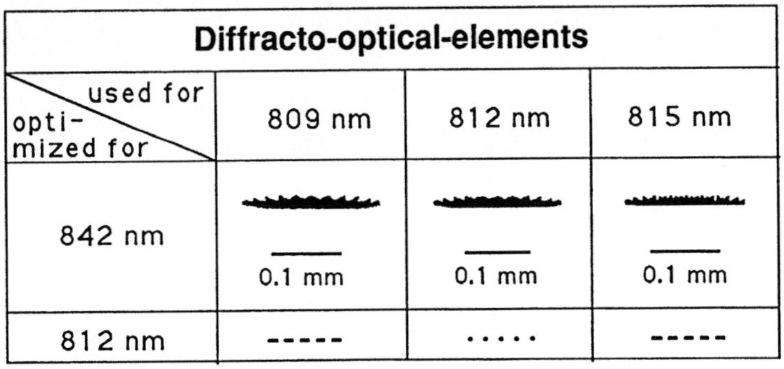

Diffracto-optical-elements			
used for / opti-mized for	809 nm	812 nm	815 nm
842 nm	0.1 mm	0.1 mm	0.1 mm
812 nm	- - - - -	. . . . .	- - - - -

Fig. 4: a) Spot-diagram of one laserpoint imaged to itself
b) Spot diagram of five laserpoints imaged to itselves

References

/1/ K. Petermann: "Laser diode modulation and noise"; Kluwer Academic Publishers, 1988

/2/ V. L. Velichanski, A. S. Zibrov et al. : Zh. Tekh. Fiz. Lett. Vol. 4 (1978) p. 1087

/3/ David Mehuys, David F. Welch, Leo Hollberg, Don R. Scifres: "1.0 W CW diffraction- limited external-cavity tunable diode laser"; Baltimore, May 1993, Conf. on Lasers and Electro-Optics (post deadline-paper)

/4/ Manfred Rothhardt, Hartmut G. Hänsel, Hans-Jürgen Dobschal: "Anordnung zur Rückkopplung von Laserdiodenstrahlung", DEOS 40 05 247

/5/ J. Sigg: "Untersuchungen zum Einfluß von externer Rückkopplung auf die spektralen Eigenschaften von Halbleiterlasern"; Dissertation, Suttgart 1987

High-Throughput Narrowband Spectral Filtering System for Raman Lidars

K. A. Stankov
Laser Laboratorium Göttingen, D-37077, Germany

Abstract:

Unconventional high-throughput multichannel narrowband spectral filtering system based on a combination of Lyot filters and Fabry-Perot interferometer was developed. The system was successfully employed in test lidar measurements.

We report on a new approach to the problem of spectral filtering in Raman lidars by using a combination of Lyot and Fabry-Perot filters. Besides the high throughput and high rejection ratio for the laser wavelength, the system features also multichannel operation which is imperative for lidar measurements with reference channels.

The system is based on a composition of Lyot filters which provide 10^9 attenuation for the fundamental wavelength 308 nm, 10^8 for the N_2 line and 10^6 for the O_2 line, as shown in Fig.1. Further x1000 suppression for the fundamental radiation as well as of the N_2 and O_2 lines is provided by a specially designed UV-Fabry-Perot interferometer with a transmission of 50% and by interference filter. Thus the total attenuation for the laser wavelength is more than 10^{15}, which exceeds that of conventional filtering systems in the UV by several orders of magnitude. The bandwidth of the system was determined by the Fabry-Perot interferometer and was matched to the 3 pm laser bandwidth.

The system was tested in laboratory conditions using EMG 160 MSC (Lambda Physics) narrowband excimer laser. The back-scattered light was detected in two channels (signal, CO_2, and reference, O_2) by means of low noise photomultipliers and photon counting technique. Excellent signal-to-noise ratio was obtained despite of the large scattering due to

TP1
Ø308
TP2
Ø308
TP3
Ø308
TP4
Ø323
GP1
Ø323
GP2
FP
L
DUG11
PM (CO2)
IF
L
DUG11
PM (O2)
IF
IF
L
DUG11
PM (N2)

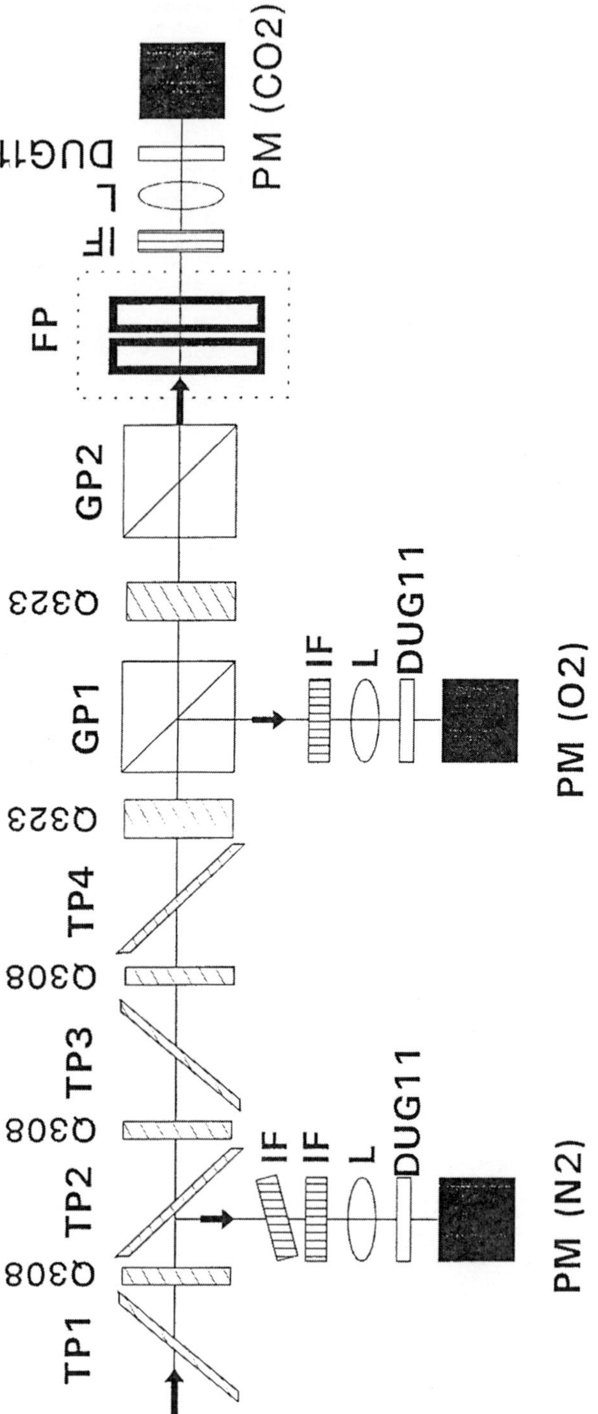

Fig. 1.

the cell windows and steering optics.

The optimum number of the thin-film polarizers in the construction of the Lyot-filters was determined to be 2, which provided a contrast ratio > 5000 At the same time, the losses arising from the thin-film polarises were kept as small as 3% per stage However, the spectral impurity of the laser radiation has limited the practical contrast and attenuation to 1000 per stage /1/. Preliminary experiments for improving the spectral quality of the commercial excimer laser (Lambda-Physics EMG-160 MSC) were executed. They were aimed at increasing the locking efficiency of the amplifier.

The final tuning of the optical filtering system was accomplished in two ways. First, in order to obtain radiation at the Raman line of the CO_2, stimulated Raman scattering was obtained using a 60 cm long cell filled with CO_2 at 10 bars pressure. This method is very convenient, since it provides high-intensity source for easy adjustment of the filter chain for the signal wavelength. As an alternative, the light from Perkin-Elmer spectrophotometer Lambda 19 was used as a high-precision light source at various wavelengths of interest (323 nm O_2 line, 334 nm N_2 line and the 321 nm CO_2 line). In this way a cross-check of the adjustment of the optical filtering system was done.

The detection of the single-photon signals from the photomultipliers has encountered difficulties due to electromagnetic interference. In order to obtain signal levels required by the two-channel photon counter, the signals need amplification by a factor of 10 or more. The four-channel high-frequency amplifier SR-445 from Stanford Research is extremely sensitive with respect to the electro-magnetic interference originating from the excimer laser. To avoid this problem, all signal cable were made with double shielding and the spectral filtering system together with the photomultipliers was enclosed in a metal box. All electronic instruments were connected with the mains with low-pass electrical filters. With these measures, satisfactorily operation was achieved.

We have performed simulation experiments which demonstrate the performance of the spectral filtering system. The experimental set-up is shown in Fig.2. The 308 nm radiation of the narrowband excimer laser EMG-160 MSC (Lambda-Physik) was directed by means of a

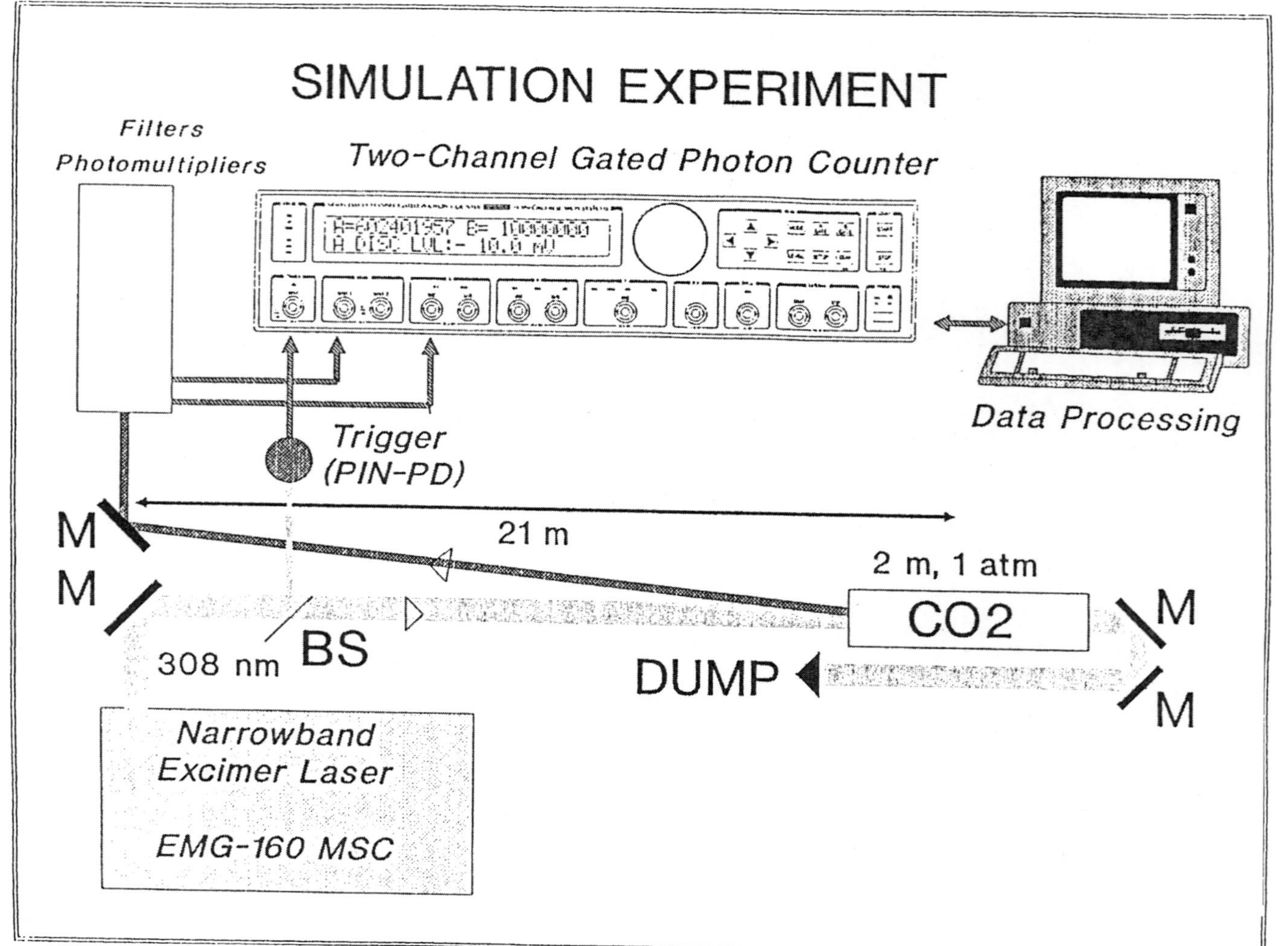

Fig. 2 . Set-up for the laboratory lidar simulation experiment.

couple of steering mirrors to a cell, situated 21 m away from the laser and the detecting system. The cell was 2 m long and filled with CO_2 gas at atmospheric pressure. Tilted quartz windows enclosed the cell from both ends. The radiation exiting the cell was properly damped. A pilot He-Ne laser beam was used to align the whole system. The back scattered light was received by the spectral filtering system, positioned at 21 m distance from the cell, on the side of the excimer laser, as is shown in the Figure. It is noteworthy that we did not use any light-collecting optics. The good signal-to-noise ratio is a proof for the high suppression of the primary laser radiation by the chain of Lyot filters.

The spectral filtering system was operated in two-channel mode. In addition to the signal channel at 321 nm (CO_2 Raman line), a reference channel at 323 nm (O_2 line) was used. As will be shown below, the availability of a reference channel is of paramount importance for the operation of the Raman lidar.

After x25 amplification by the 300-MHz SR-445 amplifier, the single-photon signals were fed into the two-channel 200 MHz photon counter (SR465 Model, Stanford Research). The latter was controlled by a personal computer.

We have performed a number of tests measurements with various repetition rates, gates, and event averaging. An example of these measurements is shown in Fig. 3. These demonstrate unambiguously the importance of the reference channel. Fig.3a illustrates the signal received by the CO_2 channel. The signal spans over considerable time delay, due to luminescence arising from the cell windows and walls as well as from the laser beam dump. The strong luminescence gives an apparent signal which looks like a signal arriving after scattering from a more distant objects /2-4/. This apparent delay is attributed to the long-lived luminescence. The situation is similar also with the other channel, tuned at 232 nm (O_2 line), Fig.3b. Only after normalisation of the CO_2 signal to that of the O_2 channel, the signals arising from luminescence in the two channels are cancelled. A well defined signal peak corresponding to the exact position of the CO_2-filled cell is observed, Fig.3c. Note the good signal/noise ratio, obtained without any light-collecting optics.

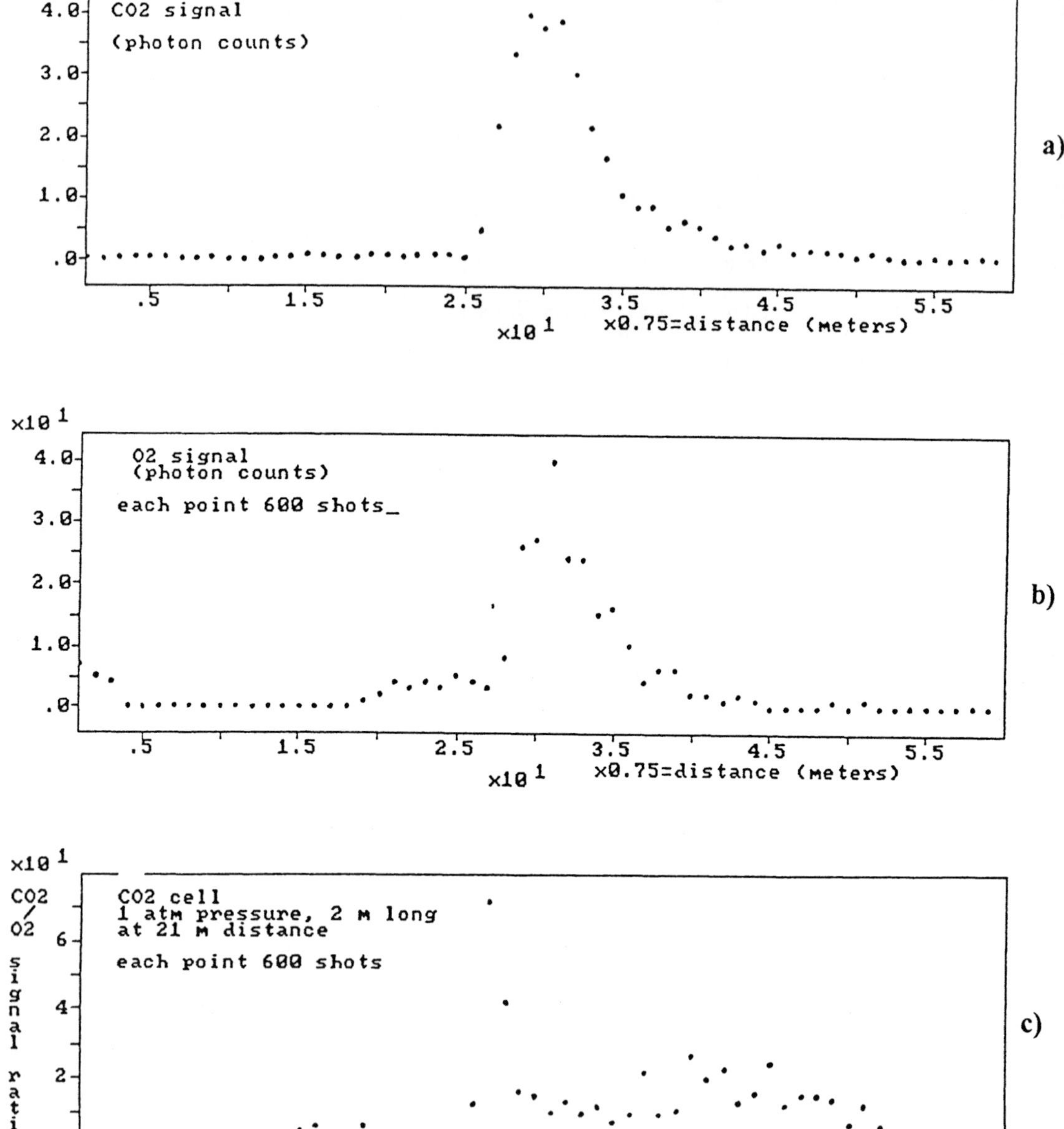

Fig. 3   a) CO_2 signal as a function of the distance from the receiver; b) the same for the O_2 channel; c) CO_2 signal normalized to the O_2 channel. The sharp peak indicates the position of the CO_2 cell.

In conclusion, a spectral filtering system suitable for Raman lidars was successfully tested. It features high transmission of 12%, and very high attenuation for the primary (laser) wavelength, reaching 10^{14} in the full version, including the Fabry-Perot interferometer. At the same time, narrowband operation of 3 pm can be achieved, which assures that the luminescence arising from tropospheric aerosols will be adequately suppressed. The laboratory simulation measurements has proved the performance of the spectral filtering system.

REFERENCES:

1. K.A.Stankov, 1991 Annual Report, Laser Laboratorium Göttingen

2. C.Weitkamp, M.Riebesell, E.Voss, W.Lahmann and W.Michaelis, Report GKSS 86/E/58

3. M.Riebesell, Ph.D. Thesis, Hamburg, 1990

4. Laser monitoring of the atmosphere, ed. E.D.Hinkley, Topics in Appl.Phys. Springer-Verlag, Berlin Heidelberg New York, 1976.

5. K.A.Stankov, in preparation.

Ti:Sapphire Based Lidar Systems

J.P. Wolf, J.Kolenda, P. Rairoux, J. Reif, M. Douard, M. Ulbricht

Elight Laser Systems GmbH; Potsdamer Straße 18A
14513 Teltow / Berlin (Germany)
Freie Universität Berlin; Institut für Experimentalphysik
Arnimallee 14 ; 14195 Berlin (Germany)
Univ. Lyon I; LASIM; 43, Bd 11 Novembre 1918
69622 Villeurbanne Cedex (France)

Recent progress in Lidar/DIAL technology has allowed to obtain 3-dimensional mappings of the concentration of air pollutants at highest sensitivity (ppb-range) and over large distances (10 km) [1-6]. Presently, it is possible to monitor real-time distributions and dynamics of nitrogen oxides, sulfur dioxide, and ozone. Recently, also the detection of toluene and benzene in the near UV has been demonstrated [7]. Routine or fully automatic operation has been severely restricted, however, by the complexity and maintenance of the usually employed Nd:YAG or Excimer-pumped dye lasers. The advent of new *tunable* all-solid-state laser systems, such as vibronic lasers (Ti:Sapphire, LICAF, LISAF,...) or laser-pumped OPOs, opens a new era in the domain of userfriendly and fully automatic DIAL operation.

Here, we present the first Lidar/DIAL systems which are based on this new type of lasers. For this purpose, a new flashlamp-pumped Ti:Sapphire laser has been developed, combining the wide tunability of the laser medium with the ease of operation, characteristic for flashlamp-pumped solid state lasers. The wavelength range accessible by the laser and its extension by nonlinear optical devices makes it an ideal tool for both DIAL and meteorologic applications.

The Laser

The short pulses with narrow linewidth and high pulse energy, indispensable for high selectivity and long detection range, are provided by an oscillator/amplifier configuration of the flashlamp-pumped Ti:Sapphire laser [8,9], with the peculiar highlight of a patented double oscillator for easy alternation or simultaneous operation of signal and reference wavelengths (see Figure 1), with a collinearity between both beams of better than 100 µrad.

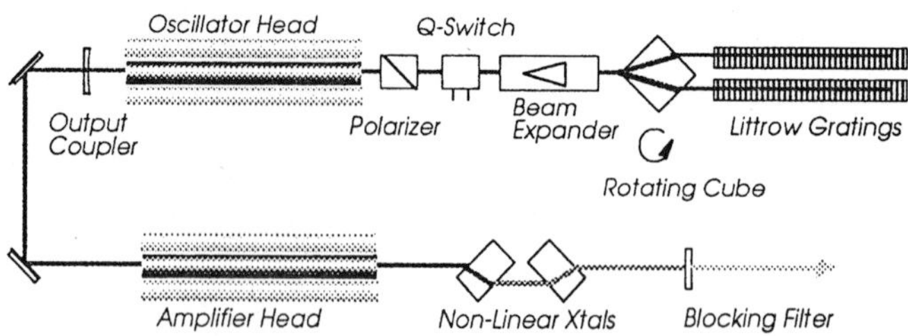

Figure 1: Flashlamp-pumped Ti:Sapphire laser for DIAL applications.

The double oscillator is tuned by two identical gratings in Littrow configuration. In the Q-switched mode, it delivers 30 ns pulses with a bandwidth of 0.2 cm⁻¹. With a 50 % output coupler and a ø 8 x 200 mm rod, a maximum pulse energy of 250 mJ is achieved. The Q-switch efficiency is then > 60 % as compared to the free running laser, and the threshold pump energy amounts to 45 J. In Figure 2, the tuning range of this laser is shown.

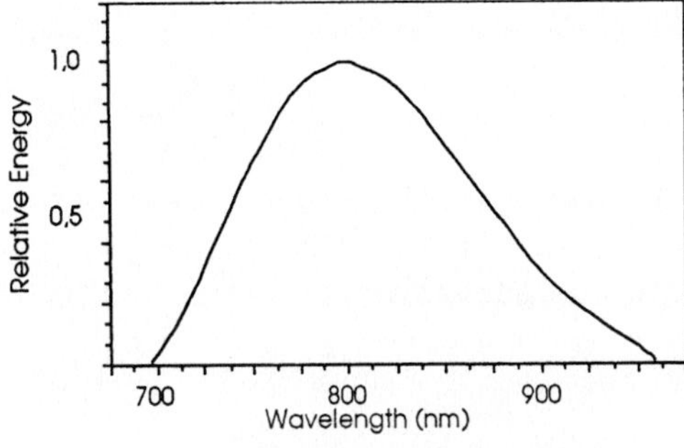

Figure 2: Tuning range of the flashlamp-pumped Ti:Sapphire laser

Figure 3: Efficiency of the Ti:Sapphire amplifier for two different rods

For the amplifier, two different rods were tested: (1) ø 8 x 150 mm; 0.15 % Ti, FOM 200; (2) ø 8 x 200 mm; 0.1 % Ti, FOM 400. In both cases, the flashlamp energy did not exceed 2.5 J/cm$_{arc}$ for flashlamp lifetime reasons. The result is presented in Figure 3. At low input energy, an amplification by a factor of ~7 is achieved with the 200 mm rod, decreasing to a factor of ~3.5 at the highest input, due to beginning saturation.

From these experiments, we expect the final specifications of the complete system as listed in the following table (@ 790 nm):

Linewidth	0. 2 cm⁻¹
Pulse Duration	30 ns
Pulse Energy	800 mJ
Divergence	1 mrad
Repetition Rate	20 Hz

Because of its high intensity at narrow bandwidth and low divergence, the laser allows efficient nonlinear optical frequency conversion (SHG ~ 20 %, THG ~ 4% of the fundamental) to reach the wavelengths which are interesting for DIAL applications.

NCPM-OPO

An additional extension of the accessible wavelength range, in particular in view of hydrocarbon detection, is provided by a KTP optical parametric oscillator (OPO) in NCPM (non-critical phase matching) configuration (Figure 4).

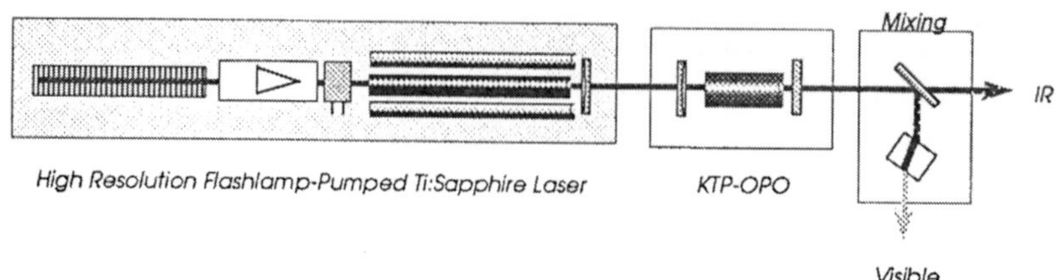

Figure 4: Wavelength extension by NTCM-OPO and frequency mixing

The big advantage of this set up is its simplicity and the fact that the OPO is free of tracking problems. Because of the tunable pump source, tuning of the OPO is possible without rotation of the nonlinear crystal. Consequently, non-critical 90° phase matching can be achieved.The diffraction of pump beam and resonated OPO signal beam upon entering the crystal is then identical, avoiding a walkoff between the two beams in the crystal. This results in a good efficiency over the whole crystal length. With this system, infrared radiation in the 1.01 - 1.43 µm and 2.16 - 3.31 µm bands is generated with an efficiency of up to 10 % and a bandwidth of ~ 1 cm^{-1}. Mixing of the OPO signal resp. idler beam with the Ti:Sapphire pump laser generates light in the visible, thus closing the gap between Ti:Sapphire fundamental and SHG. Operation of the OPO at an angle of 60° results in a further increase of the tuning range (Figure 5), at the expense, however, of introducing walkoff-problems.

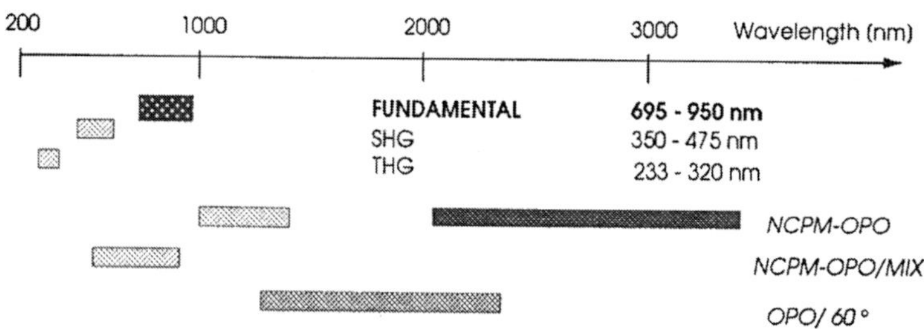

Figure 5: Tuning range of the Ti:Sapphire laser and nonlinear optical extensions

Dial Monitoring of Air Pollution

This laser system is being implemented in the already operating Lidar station over the city of Leipzig. This station is equipped with a 600 mm Cassegrainian receiving telescope and a 1350 x 645 mm turning mirror to define the observation direction. A sophisticated software package on a DEC workstation allows automatic operation and comfortable on-line data evaluation and presentation.
A smaller laser, only consisting of a power oscillator, will be the source of a mobile Lidar device on the basis of a Volkswagen van, which is presently under construction.

In order to exploit the specific tuning range of the Ti:Sapphire laser system in the most effective way, the following pairs of probe and reference wavelength have been chosen for the DIAL applications:

POLLUTANT	λ_{on} (nm)	λ_{off} (nm)	$\Delta\kappa$ (cm^{-1}atm^{-1})	SOURCE	PULSE ENERGY (mJ)
SO2	286.55	286.13	9.05	THG	10
O3 *	279.20	291.30	86	THG	10
NO	226.80	226.83	105	FHG	1
NO2 **	398.29	397.50	4.5	SHG	160
Toluene *	266.90	266.10	29.5	THG	32
Benzene *	252.90	251.95	60.8	THG	30
Methane *	2367.0	2355.0	6.5	OPO	20
CO	2347.0	2355.0	2	OPO	20
CO2	2768.0	2775.0	27	OPO	30

* Maximal difference between probe and reference for largest $\Delta\kappa$; to be fitted with concentration.
** Dissociation threshold $\approx$ 397 nm

A particulary interesting perspective for simultaneous detection and monitoring of $NO/NO_2/O_3$, relevant for summer smog situations, is opened if the laser oscillator is optimized, using specially adapted mirrors, for operation around 900 nm. Similar to the previously developed NO/NO_2 scheme [10], the following set of wavelengths can be used:

POLLUTANT	λ_{on} (nm)	λ_{off} (nm)	SOURCE
NO	226.8	224.05	FHG
NO$_2$	448.1	453.6	SHG
O$_3$	298.7	302.4	THG

Since the two wavelength triples 224.05/448.1/298.7 nm and 226.8/453.6/302.4 nm are based each on one single fundamental Ti:Sapphire wavelength (896.2nm resp. 907.2 nm), they are emitted simultaneously, and the received backscattered signals may be separated by appropriate filtering. From such measurement, the oxidation process responsible for ozone smog could be directly monitored.

Meteorologic Applications

The emission range of the Ti:Sapphire laser is also extraordinarily well adapted for the Lidar measurement of meteorological parameters.
The content of water vapor in the upper troposphere and lower stratosphere is of most important influence on the radiation transfer to and from the earth. It is a clue

to cirrus formation as well as polar stratospheric clouds (PSCs). For the investigation of H_2O concentration, three bands around 720 nm, 830 nm, and 935 nm are accessible. Because of different absorption cross sections, they permit to obtain information about different altitudes. Because of its large absorption ($\sigma_A \approx 1.88 \times 10^{-21}$ cm^2), the long wavelength band is predominantly suited for mesurements from high altitudes, e.g., airborne or based on high mountains, such as the Jungfraujoch, both projected in the near future. The simulation in Figure 6 shows the advantage of the long wavelength at high altitudes, as compared to the 720 nm band: above 2000 m, the contrast between probe and reference is almost two orders of magnitude better.

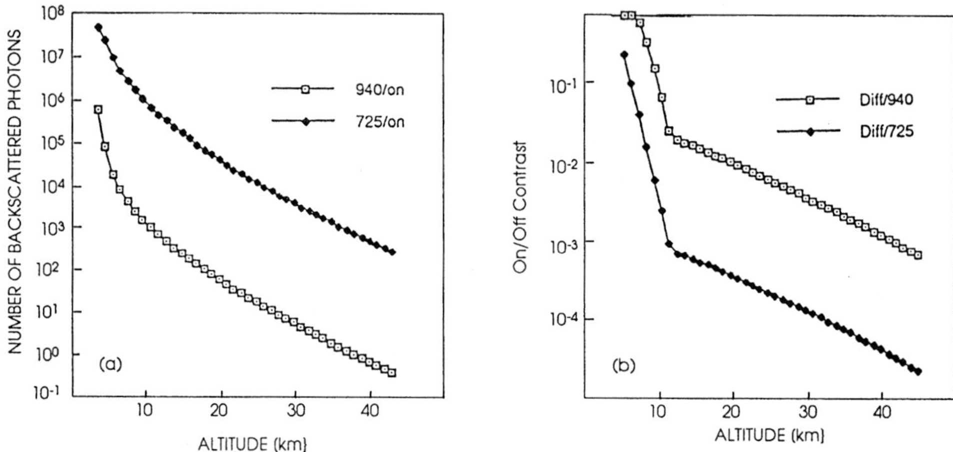

Figur 6: Lidar studies of stratospheric water concentration. (a) Atmospheric backscattering for 725/940 nm; (b) Contrast probe/reference for 725/940 nm.

The potential of the Ti:Sapphire laser for the measurement of aerosol contributions has been impressively demonstrated during the International Arctic Ozone Campaign (EASOE) in the winter 1991/92. By the combination of several wavelengths it was possible to deconvolve the particle size distributions, and thus obtain information about the dynamics of Polar Stratospheric Clouds and cirrus clouds [11]. From the different response to the different wavelengths shown in Figure 7, two types of aerosols could be distinguished. Up to 21 km, sulfuric acid aerosols of 60 % to 70 % H_2SO_4/H_2O were found, while at 21.5 km and at 22.5 km, the mean radius of the particles is significantly smaller, with the higher refractive index pointing to frozen NAT [12]. These second aerosol layers were found to move and change considerably faster than the aerosols below. The high layers could only be observed in airmasses having passed the Norwegian mountain ridge. All observations indicate, that these are in fact PSCs formed by condensation and freezing of NAT when temperature drops rapidly in lee waves over northern Norway.

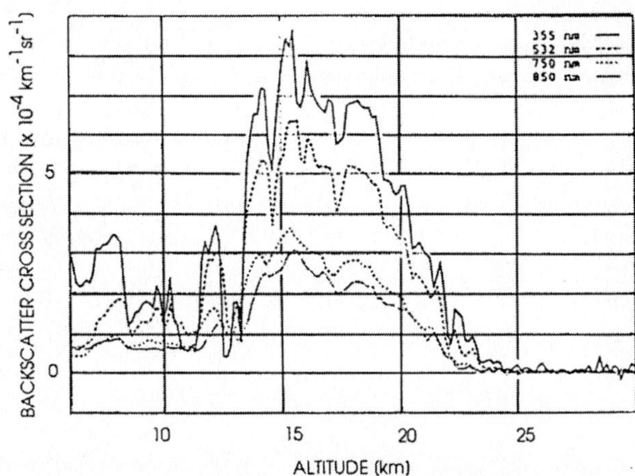

Figure 7: Backscatter cross sections from the arctic sky, due to Mie scattering for four wavelengths.

References:

[1] Weber, K., V. Klein, W. Diehl: "Optische Fernmeßverfahren zur Bestimmung gasförmiger Luftschadstoffe in der Troposphäre; VDI Berichte **838**, 201 (1990)

[2] Werner, C., V. Klein, K. Weber (eds.): *Laser in der Umweltmeßtechnik - Laser in Remote Sensing*; Springer, Heidelberg 1992

[3] Measures, R.M.: *Laser Remote Chemical Analysis*; Wiley, New York 1988

[4] Kölsch, H.J., P. Rairoux, J.P. Wolf, and L. Wöste: "Comparative Study of Nitric Oxide Immission in the Cities of Lyon, Geneva, and Stuttgart Using a Mobile Differential Absorption LIDAR System; Appl. Phys. B **54**, 89 (1992)

[5] Rairoux, P.: "Mésures par Lidar de la pollution atmosphérique et des paramètres météorologiques"; PhD Thesis, Thèse 955, EPFL Lausanne, 1990

[6] Edner, H., A. Sunesson, S. Svanberg: "NO plume mapping by laser radar techniques"; Opt.Lett. **13**, 704 (1988)

[7] Milton, M.J.T., P.T. Woods, B.W. Jolliffe, N.R.W. Swann, and T.J. McIlveen; "Measurements of Toluene and Other Aromatic Hydrocarbons by Differential-Absorption LIDAR in the Near-Ultraviolet"; Appl.Phys. B **55**, 41 (1992)

[8] Erickson, E.G.:"Flashlamp pumps Ti:Sapphire laser"; Laser Focus World **25/8**, 21 (1989)

[9] Kolenda, J.: "Anwendungen des blitzlampengepumpten Titan:Saphir Lasers in der Lidar-Technik; PhD Thesis, Freie Universität Berlin, 1993

[10] Kölsch, H.J., P. Rairoux, J.P. Wolf, and L. Wöste; "Simultaneous NO and NO2 DIAL Measurement using BBO Crystals"; Appl. Opt. **28**, 205 (1989)

[11] Kolenda, J., B. Mielke, P. Rairoux, B. Stein, D. Weidauer, J.P. Wolf, L. Wöste, F. Castagnoli, M. DelGuasta, M. Morandi, V.M. Sacco, L. Stefanutti, V. Venturi, and L. Zuccagnoli: "Aerosol size distribution measurements using a multispectral Lidar-system", SPIE **1714**, 209 (1992)

[12] Toon, O.W., E.V. Browell, S. Kinne, J. Jordan: "An Analysis of Lidar Observations of Polar Stratospheric Clouds"; Geophys.Res.L. **17**, 393 (1990)

Aerosol Backscatter Measurements Using a Compact CO$_2$ Lidar Sensor

Robert Lange, Michael Fiedler, Erich Golusda, and Klaus Lühmann
Battelle - Institut e. V.
Am Römerhof 35, 6000 Frankfurt am Main 90, Germany

There is an increasing need for air monitoring regarding hazardous organic gases which - if released into the atmosphere - have impact on the environment, on human health or might cause hazardous industrial accidents. Conventional air monitoring techniques detect only nearby pollutants and thus yield limited information on the pollutants' local distribution. As a consequence concentrations with strong spatial variations may not be detected at all or merely as apparently diluted and small. Therefore, a pronounced progress in air monitoring is expected from the DIAL technique (DIfferential Absorption LIDAR, LIDAR: LIght Detection And Ranging), since DIAL allows to identify, quantify and localize remote atmospheric pollutants from a single sensor location (without reflecting or backscattering topographic target); concentration profiles and three-dimensional maps obtained by DIAL display spatial concentration variations as well as temporal fluctuations and concentration build ups. DIAL is based on absorption spectroscopy. Since relevant hazardous gases show characteristic absorption lines which coincide with CO₂ laser lines within the 9 to 11 μm region[1], CO₂ laser based DIAL sensors seem to be well suited instruments for monitoring the atmosphere.

The sensitivity of DIAL sensing is related to the value of the (differential) optical absorption coefficient $\wedge\alpha$, the higher the absorption coefficient the lower the detectable minimum concentration n_{min} of the respective gas. Thus, the sensitivity can be expressed (independently of the respective gas) in terms

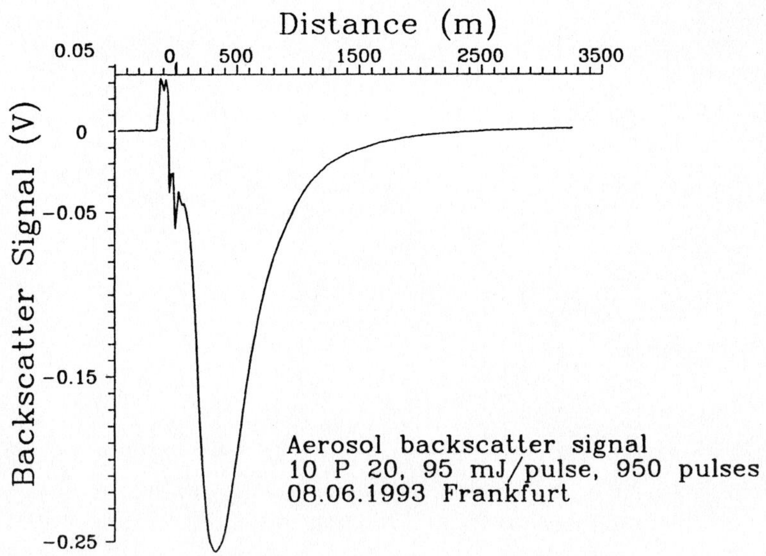

Fig. 1: Backscatter LIDAR signal of radiation scattered at atmospheric aerosols.

Below 400 m, the signal decreases with decreasing range. This effect, called (geometric[4]) signal compression, results from the the limited field-of-view which - in this range - is smaller than the diameter of the emitted laser beam. Signal compression lowers the dynamic range of the backscatter signal and thus allows to reduce the impact of the digitizer's finite dynamic resolution on the signal-to-noise ratio (especially of signals received from a large distance). Beyond 500 m the signal decreases with increasing range. The decrease is caused by the angular aperture of the receiver, by extinction and by a small misalignment between the receiver and transmitter axis.

At a distance of 500 m the electrical signal corresponding to the backscattered light intensity amounts to 0.2 V. Fig. 2 shows the same signal as Fig. 1 but with the ordinate axis stretched to read the electrical noise; it amounts to ± 0.1 mV. Thus, in case of a distance of 500 m we achieve an electrical signal-to-noise ratio of 2000.

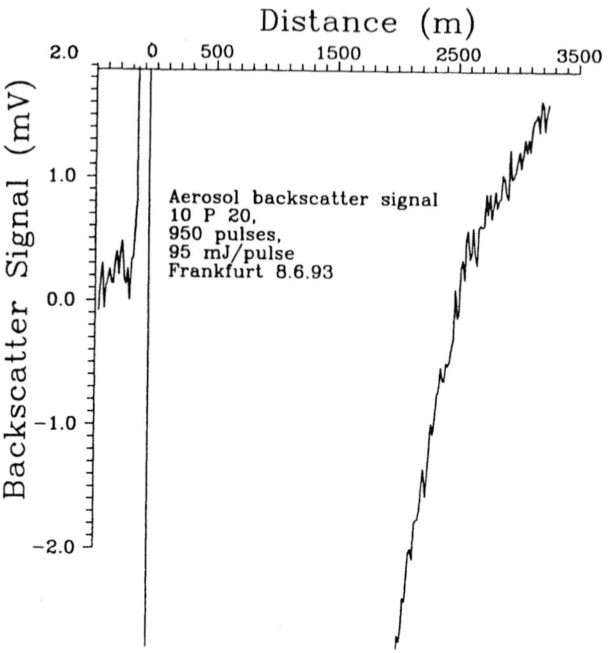

Fig. 2: Backscatter LIDAR signal of Fig. 1 but ordinate axis stretched.

Fig. 2 shows the entire range of the LIDAR Sensor which significantly is larger than 2.5 km. The backscatter signal received from aerosols in a distance of 2 km amounts to 4 mV which results in a signal-to-noise ratio of 40. As derived above this signal-to-noise ratio is sufficient for environmental and industrial applications.

Furthermore, Fig. 2 shows a shift of the "zero voltage level" resulting in a positive voltage signal of radiation scattered at a distance larger than 2500 m. This effect is due to the cut off of the preamplifier's high pass filter behaviour, with a cut off frequency estimated to some hundred Hertz. Its impact on the concentration measurement has been investigated in detail by computer simulations and needs to be compensated by appropriate DIAL evaluation algorithms.

of the (minimum) optical density defined as the product of the optical absorption coefficient and concentration. For various hazardous gases of industrial and environmental interest the optical density corresponding to the maximum allowed concentration at working places - a typical concentration to be detected by DIAL - is larger than 10^{-5} per optical pathlength through the contamination cloud[1]. Computer simulations have shown[2] that this optical density can be measured with sufficient accuracy using compact low energy lasers. To achieve this sensitivity the backscatter signal needs to be detected with an electrical signal-to-noise ratio SNR better than 30 if the absorption path length $\wedge l$ amounts to 30 m, in general

$$SNR_{min} = (n_{min} \wedge \alpha \wedge l)^{-1}.$$

At present a compact CO_2 DIAL sensor[2] is under development at Battelle. In this publication we report on the sensor performance regarding the signal-to-noise ratio of backscatter measurements. The CO_2 laser used is tunable across 68 lines. For the experiment the 10P20 line was selected emitting an energy of 95 mJ. Light scattered back has been collected by a telescope of 40 cm diameter and focused on a PV-MCT detector. A 12 bit transient recorder of 5 MHz bandwidth recorded the resulting signal which then has been averaged over 950 laser pulses and smoothed. Transmitter and receiver have been aligned by aiming at a topographic target in 2.2 km distance and maximizing the return signal.

Fig. 1 shows a backscatter signal obtained from aerosols in the atmosphere. The backscatter signal is detected as a function of time elapsed since the emission of the laser pulse. To indicate the moment of pulse emission we detected radiation scattered within the sensor together with the atmospheric return which results in the peak of the leading edge. Disregarding effects of the laser pulsform[3] Fig. 1 thus shows the detected power of light scattered at aerosols at the respective distance from the sensor.

Electromagnetic interference only occurs during the discharge of the laser condensators. It is of short duration and thus has no impact on the measurement accuracy (at least in case of a distance larger than 100 m).

In summary, the results demonstrate that LIDAR sensors operated with low energy CO_2 lasers are well suited tools for atmospheric monitoring. The sensor detects radiation scattered at aerosols with a signal-to-noise ratio which is sufficient to measure concentrations of environmental and industrial interest up to ranges beyond 2 km. To evaluate the DIAL signal in case of large ranges, artefacts resulting from the electronics need to be compensated.

Acknowledgements

The work is commissioned and supported by the German Ministry of Research and Technology (BMFT).

[1]   Fiedler, M., Lange, R.: "Monitoring of Industrially Relevant Organic Gases by CO_2 Laser Radiation", SPIE Vol. 1716 (1992), 98
[2]   Lange, R., Fiedler, M.: "Performance of a Mobile CO_2 Laser Based DIAL Sensor for Range resolved Measurements of Organic Trace Gases", SPIE Vol. 1714 (1992), 46
[3]   Falk, F., Lange, R.: "Solution Method for the LIDAR Equation", SPIE Vol. 1714 (1992), 303
[4]   Staehr, W., Lahmann, W., Weitkamp,C.: "Range-Resolved Differential Absorption Lidar: Optimization of Range and Sensitivity", Appl. Optics 24, (1985), 1950

The Stratosphere two Years after the Pinatubo Eruption

H. Jäger
Fraunhofer-Institut für Atmosphärische Umweltforschung, IFU
Kreuzeckbahnstraße 19, 82467 Garmisch-Partenkirchen

INTRODUCTION

Long-term records of the stratospheric aerosol layer exhibit periods of low and high aerosol load. Both, the background periods and the volcanically perturbed periods, are of interest. Lidar records show that the variability of the stratospheric aerosol content spans more than two orders of magnitude. Explosive eruptions penetrating into the stratosphere can provide large signals to the atmospheric radiation budget and to atmospheric chemistry. The investigation of periods without volcanic input to the stratosphere provides information on non-volcanic sources of the stratospheric aerosol. Such sources are the diffusion of sulphurous precursor gases into the stratosphere or convective processes. In addition, sources resulting from anthropogenic activities cannot be ruled out.

The stratospheric perturbation following the eruption of the volcano El Chichón (Mexico), 1982, caused an increase in aerosol surface area which was sufficiently large for measurable ozone destruction through heterogeneous chemistry processes (HOFMANN and SOLOMON, 1989). The present perturbation caused by the violent eruption of the equatorial volcano Pinatubo (Philippines, 15.1°N, 120.4°E) on June 15, 1991, appears to exceed the El Chichón event and seems to become the largest perturbation of the stratosphere ever observed by modern in situ and remote sensing techniques.

LONG-TERM OBSERVATIONS

Laser remote sensing of the stratospheric sulphate aerosol layer by ground-based lidar began at Garmisch-Partenkirchen in 1976. Since then an almost uninterrupted record exists at this midlatitude station. Until 1990 a ruby laser was used. Since 1991 a frequency doubled Nd:YAG laser transmitting at 532 nm is in operation as the lidar emitter. Atmospheric backscattering is received by a Cassegrain telescope and recorded by photomultiplier and photon counter. Lidar data are given in Table 1. The system receives backscatter from stratospheric aerosols, namely from submicrometer H_2SO_4/H_2O droplets which photochemically form from sulfurous gases. Figure 1 shows the long-term record since 1976 at the ruby wavelength of 694 nm.

Table 1. Lidar System

Location	Transmitter	Receiver
Garmisch-Partenkirchen 47.5°N, 11.1°E	pulsed Nd:YAG laser (doubled) 532 nm, 5-7 nsec pulselength, 550 mJ/pulse, 10 Hz, 0.5 mrad divergence	Cassegrain telescope, 52 cm diameter, PMT, 200 MHz photon counter

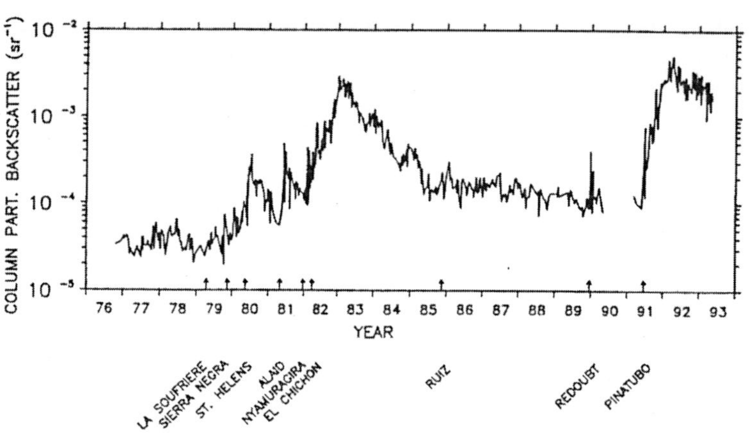

Fig. 1. Time variation of the vertically integrated particulate backscattering (integration tropopause + 1 km to layer top) at 694 nm.

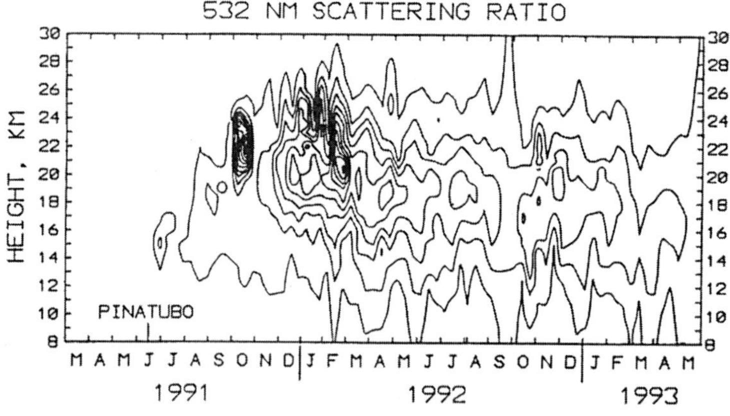

Fig. 2. Contours of the backscatter ratio at 532 nm, starting at 1.3 with increments of 0.7.

By applying a height and time resolved aerosol model, which is based on University of Wyoming balloon sonde data, the measured backscatter data can be transformed to other wavelengths and can be converted to particle extinction, mass, and surface data (JÄGER and HOFMANN, 1991).

OBSERVATION OF THE PINATUBO PERTURBATION

The first observation at Garmisch-Partenkirchen of a cloud related to the Pinatubo eruption was seen on July 1, 1991 (JÄGER, 1992). Since then the build-up to one of the largest perturbations in the stratosphere in this century was observed continuously. In Figure 2 the period 1991-93 is shown as a contour plot of the lidar backscatter ratio (ratio of measured total backscatter to calculated molecular backscatter) at 532 nm. The graph shows the early arrival, within a few weeks of the eruption, in the range below 20 km; the observation of a very dense cloud above 20 km after about 4 months during a regime of reversed summer winds (easterlies); the merging into one deep layer after the change to the winter wind regime (westerlies); the maximum appearing after 8 months; and finally a downward trend of the contours and a stabilization of the maximum at about 17 km. Figure 3 shows a Pinatubo profile from February 1992 when the load from the Pinatubo eruption was at a maximum.

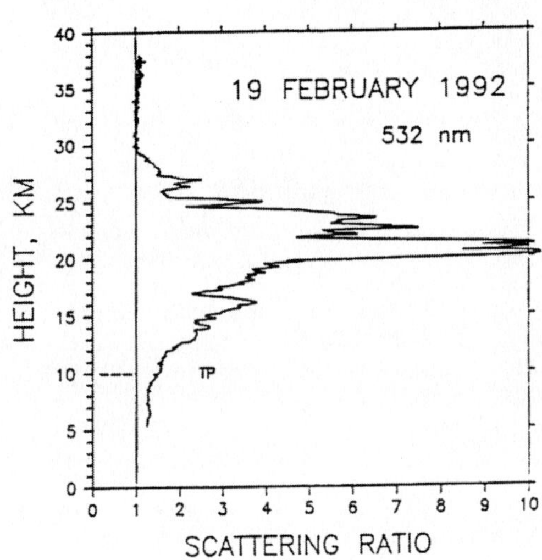

Fig. 3. Profile of the scattering ratio taken at the maximum of the Pinatubo eruption.

COMPARISON

A comparison of the Pinatubo eruption with the eruption of the Mexican volcano El Chichón in 1982 (Figure 4) shows that the Pinatubo effect exceeded the El Chichón effect in northern midlatitudes, as might be expected, in that the Pinatubo eruption injected almost 3 times more SO_2 into the stratosphere than the El Chichón eruption (BLUTH et al., 1992).

OZONE EFFECTS

Ozone depletion through heterogeneous chemistry takes place in polar stratospheric clouds, and similar reactions can occur on sulfuric acid aerosols in the midlatitude stratosphere, involving reactive compounds of anthropogenic origin (HOFMANN and SOLOMON, 1989). After the El Chichón eruption in 1982 a marked ozone deficit was observed during the winter months 1982/83 which was correlated with the maximum of the aerosol surface area in the stratosphere (JÄGER and WEGE, 1990). Very low stratospheric ozone concentrations observed in 1992 and 1993 may, therefore, be related to the precence of the Pinatubo cloud.

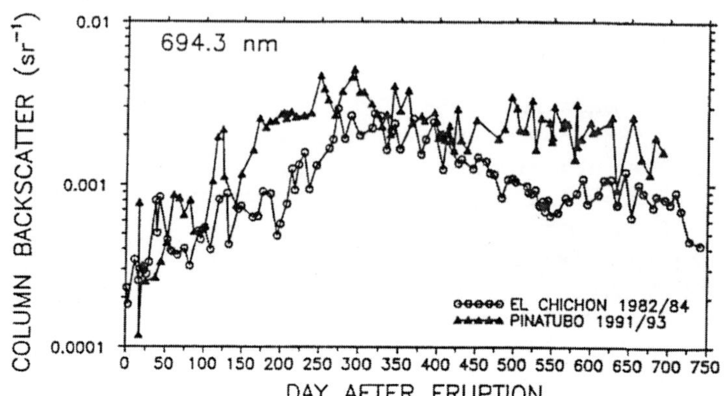

Fig. 4. Comparison of the Pinatubo and the El Chichón stratospheric perturbations at the ruby wavelength with the same integration limits as in Figure 1.

CONCLUSIONS

Lidar observations of both the equatorial eruptions of El Chichón and Pinatubo at northern midlatitudes show sofar very similar dispersion patterns. This allows to predict that the decay of the Pinatubo perturbation will still be observable until 1996. Considering the stratospheric mass loads at northern midlatitudes, the Pinatubo eruption exceeded the El Chichón eruption by a factor of 2 to 3 and the 1990 background by a factor of about 30. Extremely low ozone values presently observed at the nearby Hohenpeissenberg station of the German Weather Service indicate that volcanic aerosols seem to be involved in ozone destruction processes as it was the case after the El Chichón eruption.

REFERENCES

Bluth, G.J.S., S.D. Doiron, Ch.C. Schnetzler, A.J. Krueger, and L.S. Walter, Global tracking of the SO_2 clouds from the June 1991 Mount Pinatubo eruptions, Geophys. Res. Lett., 19, 151-154, 1992.
Hofmann, D.J. and S. Solomon, Ozone destruction through heterogeneous chemistry following the eruption of El Chichon, J. Geophys. Res. 94, 5029-5041, 1989.
Jäger, H. and K. Wege, Stratospheric ozone depletion at northern midlatitudes after major volcanic eruptions, J. Atmos. Chem. 10, 273-287, 1990.
Jäger, H. and D. Hofmann, Midlatitude lidar backscatter to mass, area, and extinction conversion model based on in situ aerosol measurements from 1980 to 1987, Appl. Opt. 30, 129-138, 1991.
Jäger, H., The Pinatubo eruption cloud observed by lidar at Garmisch-Partenkirchen, Geophys. Res. Lett., 19, 191-194, 1992.

Laser Wind Sensing: Wind Measurement by Optical Scintillation Methods

W.K. Graber and M. Furger

Paul Scherrer Intitute, CH-5232 Villigen-PSI, Switzerland

Abstract

A scintillation-based optical technique for the measurement of the average wind velocity across a laser light beam is presented and first results are shown. A low-power HeNe-laser beaming horizontally over several hundred meters through the atmosphere shows a typical optical scintillation pattern produced by atmospheric turbulence, which drifts with the transverse wind. With a simple 1-bit-correlator the laser light intensity arriving at two sensors closely spaced is analysed. From physical principles of wave propagation through a medium with refraction index irregularities the covariance of light intensity at two points can be shown to be proportional to the integral wind velocity over the light path. This instrument has the advantage over conventional wind measuring devices that it leads to an average wind information with a larger volume of representativity. Beaming across a valley, the instrument is abel to measure the total mass budget of in- and out-flow of air masses. In combination with a DOAS-system (differential optical absorption spectrometer for measuring low concentrations of atmospheric trace gases) measuring over the same path, it is also possible to calculate the mass flux of a specific species (e.g. NO_2 or O_3) perpendicular to the beam.

Introduction and outline of the method

A spatial averaging anemometer consists of a light source, beaming over a path of 0.3 to 1 km through the atmosphere. The scintillation pattern produced by drifting of refractive-index irregu-larities with the mean wind can be measured and used to calculate the average of the wind velocity perpendicular to the light path. These refractive-index irregularities are caused by atmospheric turbulence and are almost entirely the result of temperature differences. The scintillation is similar to the twinkling of stars during nighttime and can easely been observed by eye. The most effec-tive eddies, producing extinction of the light beam, are those with a half-wavelength difference in pathlength for two rays of the light source passing through the edges of the eddy. Since the light source projects the eddy on the receiver, the geometric magnification of the eddy increases towards the light source. The same increase holds for the velocity of the image drift rate at the receiver.

This phenomenon can be quantified by means of the time-lagged covariance function of the loga-rithmic amplitude of the fluctuations as given by Lawrence et al. (1972). It holds for a spherical wave with wave number $k = 2\pi/\lambda$ propagating in the z direction along the path of lenght L in the weak turbulent limit. The refractive index power spectrum of the turbulence $\Phi(K, z)$, depending on the spatial wavenumber K and the distance z from the light source, can be assumed to be linear with the refractive index coefficient C_n^2 (Kolmogorov assumption). Following these assumptions, the normalized covariance function can be calculated:

v(z) is the wind component perpendicular to the propagation of the beam, and ρ is the dis-

$$C_{\chi N}(\rho,\tau) = 2.33(kL)^{5/6} \frac{\int_0^L dz C_n^2(z) \int_0^\infty dK K^{-11/6} \sin^2[K^2 z(L-z)/(2kL)] J_0[K(\rho z/L - v(z)\tau)]}{\int_0^L dz C_n^2(z)[z(L-z)]^{5/6}}$$

placement of two sensors detecting the light beam at two neighbouring points in the receiver. A numerical simulation of this covariance function from Lawrence et al. (1972) is given in figure 1.

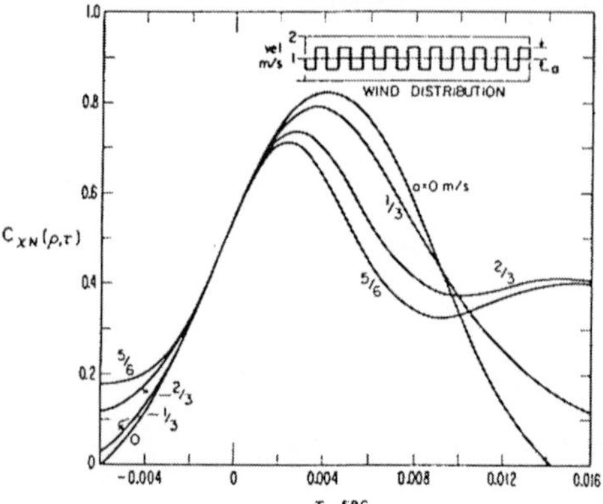

Figure 1. Time-lagged covariance functions calculated for a 1 km path and $\beta = 0.33$, using the irregular wind distribution shown in the inset. The slope of $C_{XN}(\rho,\tau)$ at zero time lag is independent of the wind distribution. Adopted from Lawrence et al. (1972).

The relation between path-averaged wind $\int v\,dz$ is rather complicated and can be linearized by differentiating with respect to τ and setting τ to zero:

$W(z)$ is the path weighting function for the slope $M_N(\rho, k, L)$ of the normalized covariance function

$$M_N = \int_0^L dz C_n^2(z)v(z)W(z) \bigg/ \int_0^L dz C_n^2(z)[z(L-z)]^{\frac{1}{3}},$$

$$W(z) = 2.33(kL)^{\frac{1}{3}} \int_0^\infty dK K^{-\frac{5}{3}} \sin^2[K^2 z(L-z)/(2kL)]J_1(K\rho z/L).$$

at zero delay. The parameter β is the the separation of the sensors in units of the first Fresnel zone, it is defined by $\beta = \rho/\sqrt{\lambda \cdot L}$.

The first Fresnel zone F_1 is defined as the distance perpendicular to the light propagation where the difference of the direct light beam L and the edge light beam L_1 is one half the wavelength: $L_1 = L+\lambda/2$ and therefore $F_1 = \sqrt{L\lambda + \lambda^2/4} = \sqrt{L \cdot \lambda}$. The condition for maximal light extinction therefore is satisfied by a Fresnel-zone size eddy of radius $r = \sqrt{\lambda \cdot z(1 - z/L)}$. Eddies of radius larger or smaller than r have focal lengths shorter or longer than L and therefore contribute less to the variance in irradiance received at a distance L from the light source.

The choice of β determines the relative weights for the wind measurement of the different portions of the path as can be seen from the numerical simulation of the weighting function in figure 2. In our instrument, the value $\beta=0.33$ is used which gives relatively uniform weighting over the path.

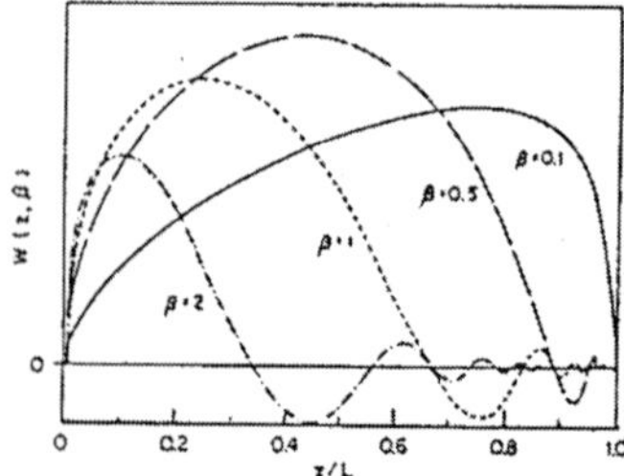

Figure 2. Path weighting function $W(z,\beta)$ as a function of the normalized distance z/L for various values of the parameter β.

The implementation of the crosswind velocity measurement is performed by means of an electronic one bit correlator. The outline follows a circuit designed by Lawrence et al. (1972) and is plotted in figure 3. The two receiving signals are logarithmically amplified and clippered (figure 4) to form a signal changing between two states. Both signals are delayed by a similar time delay determined by the two shift registers. After passing the shift register, the signal is coupled to the clippered signal of the other channel by a exclusive-or-gate (XOR). The output of an XOR is high, if one and only one of the inputs is high. Thus, the output of an XOR corresponds to the correlation function of two digital input signals, what is identical to the covariance with zero delay. The covariance with time lag τ of two digital signals is achieved by first delaying one of the signals by means of the shift register described above. To get the slope of the covariance function at zero delay, the covariances in two points have to be measured. The two points of the covariance function are determined on either side of zero delay by delaying crosswise both channels before connecting them to the input of the XOR. To change sign, the second channel after the XOR is inverted. Both channels are finally added and integrated with a time constant of 1 s. Figure 5 shows a snap shot of the succession during a wind measurement over the atmosphere.

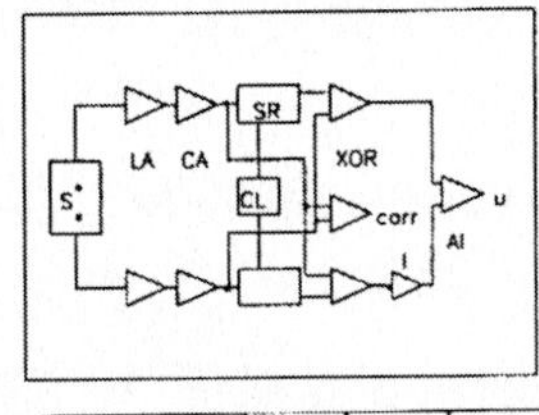

Figure 3. Electronic curcuit of the 1-bit-correlator for the slope detection of the covariance function. S=two sensors behind a narrow optical band-pass filter of 1 nm, LA=logarithmic amplifiers, CA=clipping amplifiers, CL=clock generator for the shift registers (SR), XOR (see text), I=inverter, AI=signal addition and integration (1 s time constant),u= velocity output, corr=correlation output.

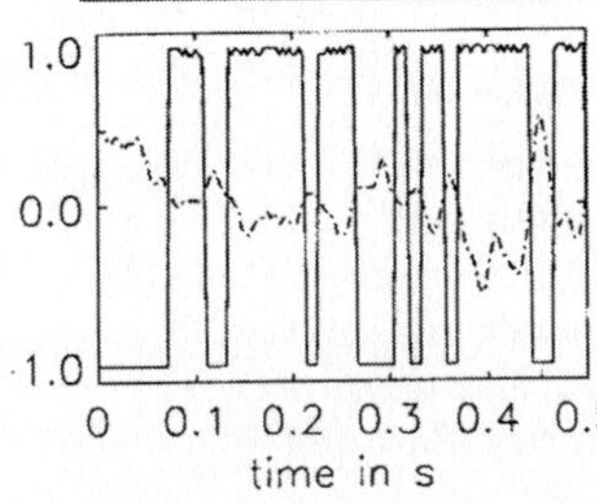

Figure 4. Time sequence of 0.5 s of the signal from one channel, after the logarithmic amplifier (dot-dashed), after the clipping amplifier (bold), scale in arbitrary units.

The choice of time delay

The time span over which the slope of the covariance function is measured is chosen to be about 3 ms. This is given by optimizing the accuracy of the wind measurement near zero and near full-scale wind speed. Near zero wind, a large time delay is optimal, since a small difference of large signals occures. On the other hand, near full scale, the signal difference is taken at points, where the linearity of the covariance function no longer holds.

It is obvious, that for two square waves with a phase difference of 90° the normalized covariance at zero delay has a slope which corresponds to one over one fourth of the time period of the square wave. In the example of figure 6 the two square waves with equal frequency of 113 Hz and a phase shift of 90° are shown. The time delay is set to $\tau=1.5269$ ms, the readout of the instrument for this setup is 5 V. It can be seen from figure 6 that the maximum output, corresponding to a maximum normalized covariance of 1, is reached for an input frequency of $f_{cal} = 1/4\tau$.

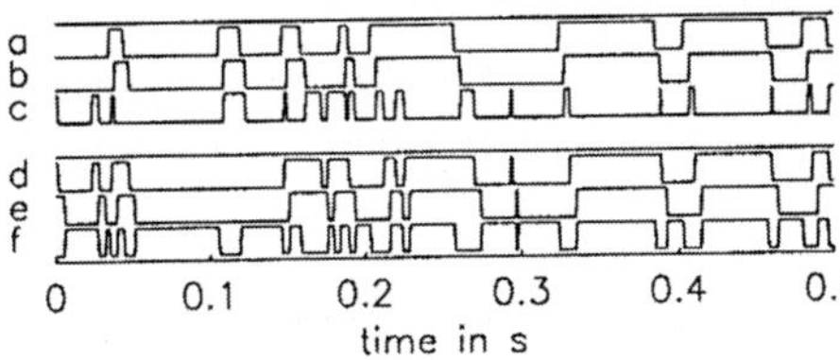

Figure 5. A snap shot of a measured time sequence of the signals at (a) the output of the first clipper, (b) the output of the first shift register, (c) the output of the first XOR. (d,e) same for the second channel, (f) the output of the inverter after the second XOR. The time delay is 1.5 ms.

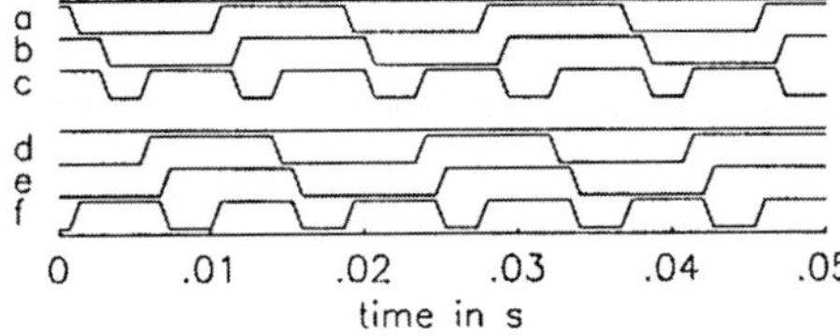

Figure 6. A time sequence with two square waves (113 Hz) with 90° phase shift connected to the input of the clippers. The points (a-f) correspond to figure 5. The time delay is 1.5 ms.

The optical setup and its calibration

The weighting function depends on the separation ρ of the two points for which the covariance function is determined. The optical sensors in our instrument have a fixed spacing ($\rho\prime = 1.3\ mm$). They are mounted behind a lense of 150 cm focal length and can be shifted along the beam. This allows the separation of the sensors to be adapted to different path length L. Our choice is $\beta = 0.33$ and therefore with the HeNe-Laser light frequency of $\lambda = 632.8\ nm$ $\rho = 8.301\ mm$. The distance b of the image behind the focal point of the lense for a light path L in km is $b = 24\ mm\,/\sqrt{L}$, consequently the sensors must be shifted to this point.

Calibration of the electronic unit

The instrument is calibrated experimentally to establish the instrument calibration factor c_0 between the output voltage of the summation-integrator and the true wind speed average over the path in the direction given by the sensor alignment. This calibration is done over a path length of 1 km and for a full scale reading of the wind speed of 10 m/s. It is essential to adapt this full scale reading to the maximum wind occuring over the path during the measurement period. To calibrate the instrument for a different path length L in km and a full scale reading of u_{max}, the calibration factor changes in the following manner: $c = c_0 \cdot \frac{\sqrt{L}}{u_{max}}$. Since the slope of the covariance function for wind velocity measurements depends on the square-root of the path length L and on the desired fullscale reading u_{max}, the time delay should be adapted to these values. The slope increases linearly with increasing full scale velocity and decreases with increasing $\sqrt{L}$, therefore the relation between the delay τ_0, for which the standard calibration factor c_0 is established and the delay τ for L and u_{max} is given by $\tau = \frac{\sqrt{L}}{u_{max}} \cdot \tau_0$. If τ is changed in this way, the calibration factor c_0 holds for all settings.

Limitations of the instrument

The measurement of crosswind velocity is not influenced by the wind along the light-path, what is in contrast to other methods as shown in a theoretical comparison of severel wind determination procedures from optical scintillation measurements by Wang et al. (1981).

The wind weighting function is valuable only under homogenious conditions of C_n^2, or the average is taken over enough time in order to guarantee statistically uniformity of C_n^2 over the path. Since turbulence is strongest at locations of maximum wind shear, not at locations of maximum wind velocity, turbulence is uncorrelated with the wind. Wang et al. (1981) show, that the deviation of the wind measurement is 27%, if one consideres a nonuniformity given by a tenfold increase of C_n^2 in 1/10 of the path, while wind velocity is assumed uniform along the path.

The instrument is not suited for measurements under conditions, where saturation of scintillation occures. This condition is given, when the atmospheric turbulence increases over a point, where the scintillation fluctuations no longer change. A critaria to decide, whether a saturation occures, is given in Ochs and Cartwright (1985). The situation is observed if the path is close to the ground and the sun heats the ground, producing strong thermal rising. The criteria is dected by measuring the correlation of channel A and B: If this correlation will become greater than 0.4, then scintillation saturation is likely to occure. The range of the instrument is limited to distances on the order of 1 km because of the increase in saturation effects with distance and the decrease in the signal-to-noise ratio to unacceptable levels. Following a formulation of the covariance function under scintillation saturation given by Ochs et al. (1976), an improved instrument with an incoherent light source and finite apertures was designed by Ochs and Wang (1978). This design is the basis for further improvements concerning the electronics with a servo loop shift register clocking and a combination of the wind measurement with direct C_n^2 measurement (see e.g. Ochs and Cartwright (1985))

A further limitation is given by the fact, that the normalized time-lagged cross correlation function for a spherical wave as given in the equation above is valid only for propagation through homogenious and isotropic weak turbulence having the Kolmogorov-type spectrum, characterized by the refractive-index structure coefficient C_n^2. The Kolmogorov model is assumed to represent turbulence in the inertial subrange, a range of eddies from a fiew millimeters to several tens of meters in size. The assumption, under which the covariance equation is derived, is Taylor's "frozen turbulence hypothesis", which states that for sufficiently short time periods the refractive-index irregularities in the atmosphere are drifting with the wind without changing its shape. This holds for weak integrated turbulence only and is specified by the the Rytov parameter σ_χ^2 not exceeding 0.3. For a spherical wave this parameter is given according to Tsadka et al. (1988) by the relation $\sigma_\chi^2 = 0.124\, C_n^2\, k^{7/6}\, L^{11/6}$. The "frozen turbulence hypothesis" limits the separation of the sensors to e few centimeters.

Future improvements

The use of uncorrelated light and larger apertures avoid the problem of the saturation scintillation as described above. Furtheron, the variation of the aperture and the use of spatial filters allow the variation of the path weighting function in a rather sharp manner and therefore the use of the light beam for ranging. This setup is termed "SCIDAR" (scintillation detection and ranging). In this setup, the spatial filter of the transmitter and receiver respectively determine the sizes of turbulent eddies that contribute the strongest signals to the receiver from each position. The relationships between zero-mean filter element size, path position, and spatial wavelength are given by Ochs et al. (1988):

$$z = \frac{L}{(1 + d_r/d_t)} \, , \; w = \frac{d_r \cdot d_t}{(d_r + d_t)}$$

with z=path position in meters, L=path length in meters, w=spatial wave length in cm, d_t=transmitter zero-mean filter diameter in cm and d_r=receiver zero-mean filter diameter in cm.

Applications and advantages of optical wind measurements by scintillation

Remote probing is especially valuable when the location at which a measurement is needed is not accessible. For instance the valley wind in an Alpine valley can be studied by crossing the valley from one slope to the other with the light path. Following a proposual by Ochs et al. (1985) for determining the refractive structure parameter C_n^2, it should be possible also for wind detection to use a folded light beam by means of a retroreflector. In this case transmitter and receiver are located at the same place and there is no power needed at the remote place. With such an arrangement wind measurements at several locations can be made by use of a single equipment mounted on a stepper motor controlled unit and pointing sequentially to several retroreflectors. This is a relatively inexpensive way to control wind velocities over large areas.

Since the scintillation anemometer measures the average wind velocity normal to a line, a mass flow perpendicular to this line is measured. In combination with a differential optical absorption spectrometer (DOAS, see e.g. Graber et al. (1992)) beaming along the same path and measuring the concentration of a gasous pollutant, it is possible to measure the true mass flow of this pollutant across the path, as demonstrated in figure 7. The measurement was made over a path of 975 m length at a height of 3 m over a plane in the "Seeland" near Berne. Small hills of 200 m height and weak slopes form the boundary of the plane. Figure 7 also shows the reading of an anemometer placed in 5 m above the ground near the receiver. Although the comparison of a point with a line measurement is restricted to homogenious wind fields, which does not hold in this case, the agreement is satisfactory. The method of determining the total mass flow of an atmospheric compound is of special interest for the determination of the pollution input from the plane into a valley by means of the valley wind, e.g. the exchange of air pollution between the polluted area of the Swiss plateau and rural sites in the Alpine valleys.

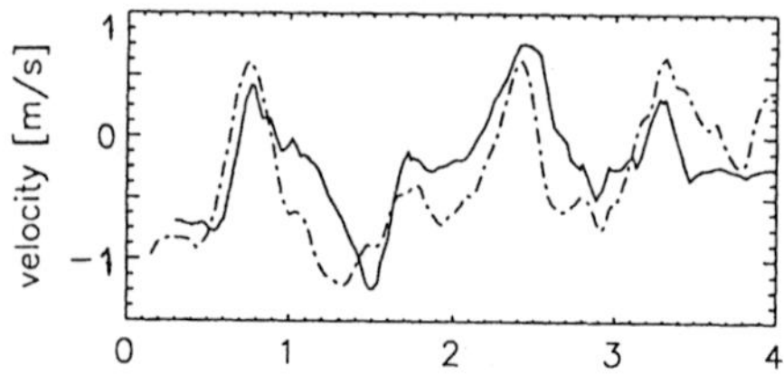 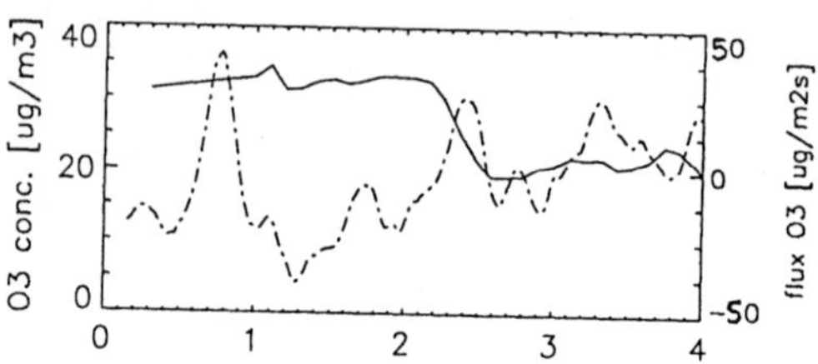

Figure 7.Left: wind station reading (bold) and Laser anemometer reading (dashed-dotted). Right: O_3 concentration measured by DOAS (bold) and O_3-flux (dashed-dotted) from 00:00 to 04:00 of June 18, 1993.

A further application of optical scintillometers is the measurement of the convergence of air masses by means of three similar optical systems forming an equilateral triangle. Such a setup is presented by Tsay et al. (1980).

Scintillation is well suited to determine the turbulent structure of the atmosphere and can be used to measure the heat fluxes near the ground as shown by Thiermann and Grassl (1992) or Hill et al. (1992). The authors make use of the relation between the refractive index coefficient C_n^2 and the temperature sturcture parameter C_T^2.

Literature

Graber W.K., Taubenberger R., Kindler T. (1992) Messung atmosphaerischer Spurengase mit der DOAS-Methode. in: C. Werner, V.Klein, K.Weber (ed.): Laser in remote sensing, 10.Int.Congr. LASER91, Muenchen, Springer 1992,210-216.

Hill R.J., Ochs G.R. and Wilson J.J. (1992): Surface-layer fluxes measured using the C_T^2-profile method. J. Atm. Ocean. Techn. ,9,526-532

Lawrence, R.S., Ochs, G.R., Clifford, S.F. (1972): Use of Scintillations to Measure Average Wind Across a Light Beam. Applied Optics 11,239-234.

Ochs, G.R., Clifford S.F. and Wang, T. (1976): Laser wind sensing: the effects of saturation of scintillation. Appl. Opt.,15,403-408

Ochs G.R. and Wang, T. (1978): Finite aperture optical scintillometer for profiling wind and C_n^2. Appl. Opt.,17,3774-3778.

Ochs, G.R. and Cartwright W.D. (revised 1985): Optical system Model IV for space-averaged wind and Cn2 measurements. NOAA TM ERL WPL-52.

Ochs, G.R., Reynolds, D.S., Zurawski, R.L. (1985): Folded-path optical Cn2 instrument. NOAA TM ERL WPL-132.

Ochs, G.R., Wilson J.J., Abbott S. and George R. (1988): Crosswind profiler model II. NOAA TM ERL WPL-152

Thiermann V. and Grassl H. (1992): The measurement of turbulent surface-layer fluxes use of bichromatic scintillation. Boundary-layer Meteor. 58,367-389

Tsadka S., Shaft S. and Azar Z. (1988): Wind velocity measurements by optical scintillations methods. SPIE,1038,548-554. Meeting in Israel on Optical Engineering(1988).

Tsay M.-K., Wang T.-I, Lawrence R.S., Ochs G.R. and Fritz, R.B. (1980): Wind velocity and convergence measurements at the Boulder atmospheric observatory using path-averaged optical wind sensors. J. Appl. Meteorol.,19,826-833.

Wang, T. Ochs G.R. and Lawrence R.S. (1981): Wind measurements by the temporal cross-correlation of the optical scintillation. Appl. Opt.,20,4073-4081

Line-Tunable Electronic-to-Vibrational Energy Transfer Lasers in the Mid-Infrared

TH. MILL, S. R. LEONE
Joint Institute for Laboratory Astrophysics, National Institute
of Standards and Technology and University of Colorado,
Boulder, Colorado, USA

Applications of laser technology like sensing and tracking of atmospheric gases require high power, line-tunable lasers in the mid-infrared, where most molecules show characteristic absorption spectra. Laser based chemical analysis in the wavelength region $2.8\mu m \leq \lambda \leq 8.5\mu m$ usually employs lead sulfide diode lasers, which are tuned by varying the diode current and diode temperature in the temperature range below 100K. However, these lasers do not achieve the efficiency and beam quality of III-V-semiconductor based diode lasers, emitting in the near infrared or visible spectral region while operating at room temperature. The possibility of collecting the output power a large number of III-V-semiconductor diodes arranged in diode arrays is now widely used for efficient pumping of high power Nd:YAG-lasers.

A way of converting conveniently available, visible wavelengths into the mid-infrared region can be provided by using either the frequency-doubled Nd:YAG-laser output or -which should become possible in the near future- the diode array light directly for photolysis of halogen atom-containing compounds e.g. I_2, Br_2. IBr or C_3F_7I. It is known that e.g. photolysis of IBr with frequency doubled light of an Nd:YAG-laser at 532 nm generates spin-orbit excited $Br(4^2P_{1/2})$ with a quantum yield of ≈ 0.7 vs. ≈ 0.3 for the generation of $Br(4^2P_{3/2})$-ground state atoms[1]. Various authors converted the inversion of the Br-atom states into laser action via the $Br(4^2P_{1/2}) - Br(4^2P_{3/2})$- transition at 2.7 μm [2,3,4], achieving pulse energies of 3 mJ [4]. This is equivalent to an efficiency of 0.08 for converting Nd:YAG-532nm-pump laser photons into Br-laser photons.

As an alternative, the available Br^*-energy can be used for vibrational excitation of various molecular species, making use of $E \rightarrow V$-transferpropensity rules [5], followed by stimulated emission via a vibrational band in the mid-infrared [4,6,7]. A CO_2-laser pumped by $E \rightarrow V$-transferfrom Br^*-atoms and emitting at 4.3 μm via the $CO_2(101) \rightarrow (100)$-band has been described [4,6,7]. Petersen and Wittig [8] demonstrated a flashlamp-pumped $Br^* \rightarrow CO_2(101)$, $E \rightarrow V$-transfer CO_2-laser with an intracavity grating for line-tuning and a partial reflector for outcoupling of the CO_2-laser pulses.

In this article, we describe a Nd:YAG laser-pumped $Br^* \rightarrow CO_2(101)$, $E \rightarrow V$-transfer CO_2-laser with two output beam portions for power output and simultaneous beam diagnostics and a cavity designed for accomodating stimulated emission arising from one or more rotational transitions of the $CO_2(101) \rightarrow (100)$- vibrational band at 4.3 μm.

The relevant energy levels of Br^* and CO_2 are shown in figure 1. After photolysis of IBr, the resulting Br^*-atoms are quenched mainly by IBr, Iodine-atoms and CO_2 with rate constants $1 \times 10^{-12} cm^3/s$ [9], $1.8 \times 10^{-11} cm^3/s$ [10], $1.36 \times 10^{-11} cm^3/s$ [11], respectively. Given these rate constants and typical pressure conditions of 267 Pa IBr and 133 Pa CO_2 in the cell, about 35% of the Br^* is quenched by IBr and Iodine atoms, while 87% of the Br^*-atoms quenched by CO_2 produce CO_2 in the vibrational state (101) [11]. Unlike the state (001), which acts as the upper laser level for the $10.6\mu m - CO_2$-laser, the

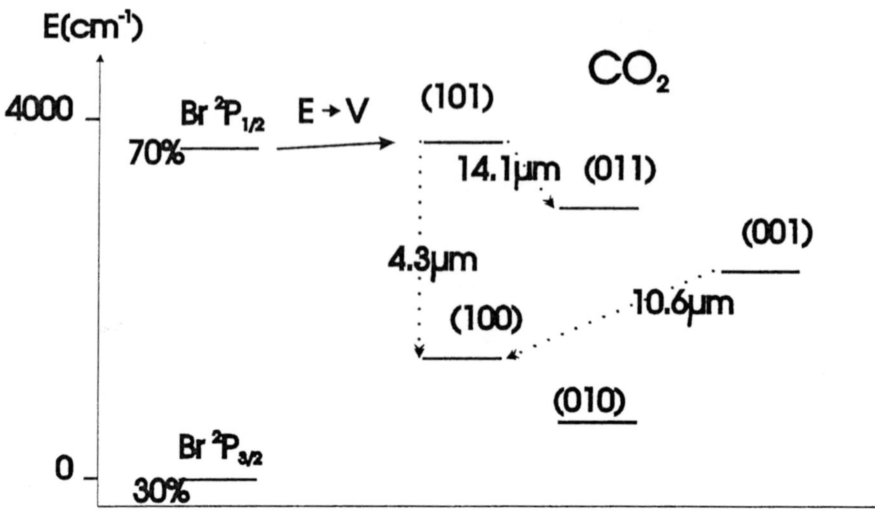

Figure 1: Energy levels of Br and CO_2

vibrational level (101) is quenched very rapidly by ground - state CO_2 (rate constant $1.16 \times 10^{-10} cm^3/s$ [11]). This limits the amount of CO_2 in the laser cell to ≈ 300 Pa for the $4.3 \mu m$ - laser, while $E{-}V$ - pumped CO_2 - lasers at 10.6 μm have been realized for CO_2 - pressures up to $\approx$ 10000 Pa [7].

The experimental setup of the authors is depicted in figure 2. A 125 cm long glass laser cell, sealed with CaF_2 Brewster - windows, is used to contain the IBr/CO_2 - mixtures. The laser can be operated with static gas mixtures, since the IBr pump material is fully regenerated after photolysis. The 2nd harmonic of a Nd:YAG - laser (532 nm, 200 mJ, 10 ns duration, doughnut - mode with 0.2 cm^2 area) photolyzes $\approx$ 25% of the IBr, given an IBr - pressure of 267 Pa. Right angle quartz prisms are used to couple into the cell at an angle $< 1^0$ with respect to the lasing axis while avoiding damage to the resonator optics.

The IR - resonator consists of plane or concave silver coated mirrors and a diffraction grating (300 grooves/mm) in Littrow configuration, reflecting $\approx$ 90% of the $CO_2(101){\rightarrow}(100)$ radiation back into the cavity while $\approx 8\%$ is deflected out of the cavity onto a gold - coated plane mirror. Both grating and mirror are mounted on the same rotatable stage with their surfaces on planes intersecting at the axis of rotation. This configuration results in an output beam that is laterally shifted by a fixed distance and paralleling the resonator mode axis, independent of the selected wavelength. In order to determine the wavelength, relative intensity and temporal behavior of the CO_2 - laser pulses obtained for various angles of the littrow grating, the reflection off the Brewster - window is observed through a 60 cm - monochromator (grating 147.5 grooves/mm) by an In:Sb - detector operating at T=77 K.

The CO_2 - laser, operating without Q - switch, produces pulses of typically 7 μs duration. With a single gas fill, operation at 10 Hz pulse frequency is stable for hours. For IBr and CO_2 used in a 2:1 - mixture, laser action is obtained for total pressures ranging from 40 Pa to 1000 Pa. Independent of the curvature of the end mirror used (R= 400 cm, 1000 cm or ∞ cm), at a total pressure of p=400 Pa, 19 rotational lines of the P - branch and 13 lines assigned to the R - branch are observed. The gain of P-branch lines is generally higher due to the degeneracies of the rotational states involved in the transition. The 32 lines observed span a tunability range of 2282 - 2356 cm^{-1}, corresponding to $\Delta\lambda$ = $0.14 \mu m$. The highest output energies of 150 μJ were observed at p=400 Pa

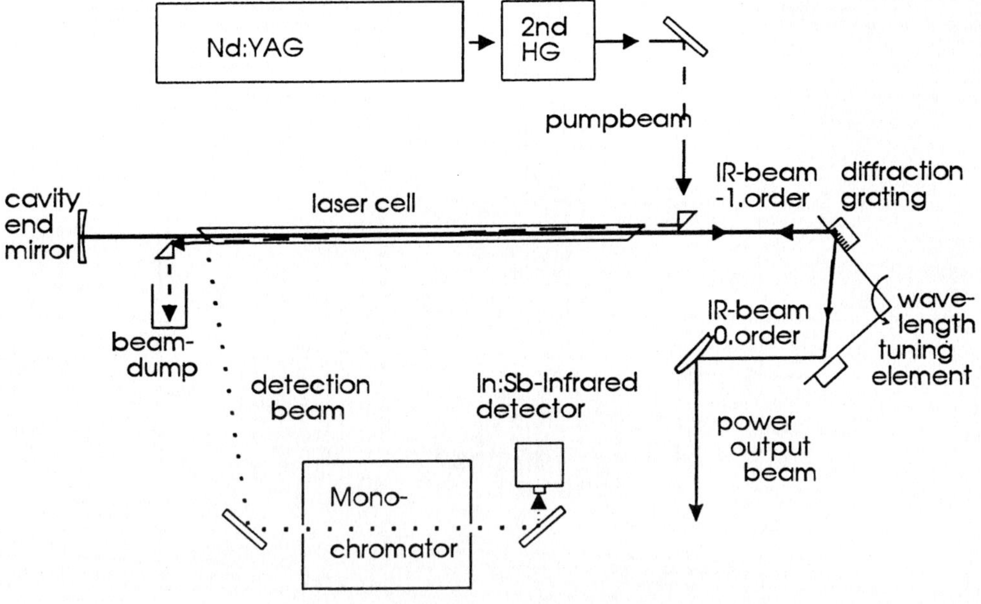

Figure 2: Experimental setup

for the lines P24, P26 and P28, which corresponds to an efficiecy of $\approx 0.6\%$ of converting the pump laser photons into 4.3 μm photons. The time interval between the pump pulse and the time when the system crosses the laser threshold is only 150 ns under this conditions. At that time only 11% of the available Br^*- energy has been transferred to the $CO_2(101)$- vibrational state. Therefore, Q- switching may be a promising way of improving on the available pulse energies.

Figure 3 shows the pulse energies of the observed P- branch lines together with a population distribution of the rotational states of $CO_2(101)$, where a rotational temperature of $T_{rot} = 300K$ is assumed. Clearly, rotational relaxation within the $CO_2(101)$- vibrational state has the effect that the pulse energies extracted from the upper laser level via rotational lines arising from states with high rotational quantum number are considerably greater than the population fractions of these states indicate, when compared with the fractions of the lines with the greatest intensity. This makes it likely that laser action can be obtained also from larger molecules with vibrations that are resonant with Br^*- or I^* - states. The number of laser lines obtained from these molecules should be larger than from the triatomic molecules CO_2 and HCN.

While the CO_2- laser output was single line under all conditions in which the flat end mirror was used, the use of concave end mirrors makes simultaneous laser emission via several rotational lines possible. Figure 4 shows how the modes arising from three rotational levels fit into the cavity and how the reflections off the Brewster- window allows them to be detected without rearranging the setup. A simultaneous lasing of up to four lines was achieved by using the R = 400 cm end mirror and of one or two lines for the R = 1000 cm end mirror. Using a convex mirror of R = 400 cm in a path distance of 200 cm from the end mirror makes the beam output from the different lines parallel, a feature that makes the CO_2- laser easier to use for analytic applications of spectroscopy. Given a defined position of the littrow grating, the number of lines lasing can be selected in this simple setup by adjusting the width of an intra- cavity aperture. The resolution of our monochromator is not good

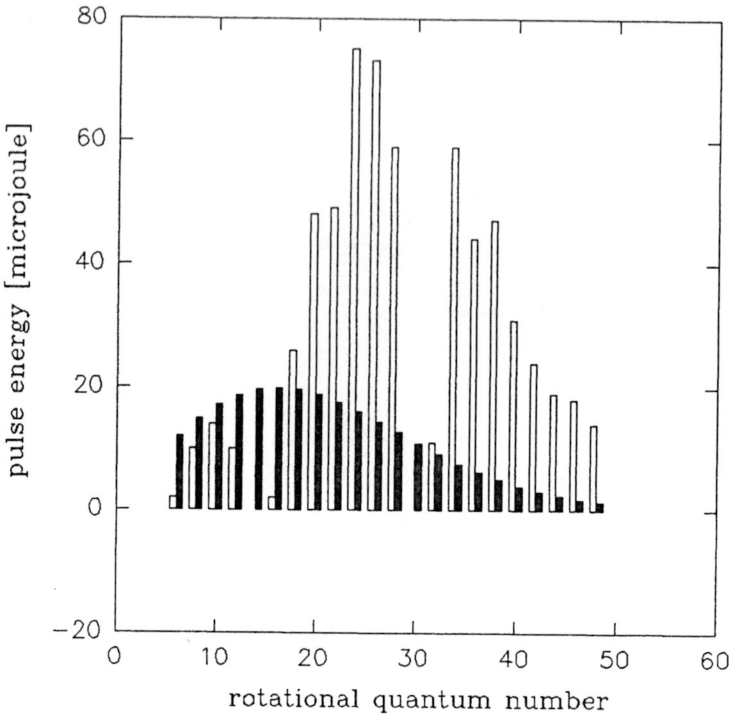

Figure 3: Pulse energies of the observed lines of the P - branch and population distribution within the $CO_2(101)$ - vibrational state at room temperature

enough to determine the spectral structure of the output of each rotational line, but considering the Doppler - linewidth of CO_2 at 4.3 μm of $\nu_D = 0.0043 cm^{-1}$ and the longitudinal mode spacing of an 100 cm long cavity of $\Delta\nu = 0.005 cm^{-1}$, single longitudinal mode operation for each rotational line should be possible. Therefore, the multi - wavelength output of the 4.3 μm CO_2 - laser is clearly defined and selected by the laser material itself and does not have to be generated by resonators selecting two discrete wavelengths from a continously emitting laser material, a feature that is required for Differential Absorption Spectroscopy in the atmosphere.

Of the rotational transitions of the $CO_2(101) \rightarrow (100)$ - vibrational transition, there are several lines missing in the output spectrum of our CO_2 - laser. Except for the line P_{30}, whose absence is due to absorption by ground - state CO_2 in the cell via the $(001) \leftarrow (000)$ - P_{52} rovibrational transition. according to the transition frequencies given by [13], the missing lines, i.e. P_{14}, R_{12}, R_{14}, R_{24}, R_{26}, R_{32}, R_{36} and R_{44} may well be due to intra - cavity absorption by gases in atmosphere of the laboratory.

Work on atmospheric constituents to be detected by the line - tunable 4.3 μm - CO_2 - laser or one of the other $E \rightarrow V$ - transfer- pumped mid - infrared lasers is still in progress. Detection of ground - state CO_2 via the $(001) \leftarrow (000)$ - vibrational transition is not possible at the present stage, since pumping with an Nd:YAG - laser at an angle to the CO_2 - laser axis does not produce an inversion of the (101) - state with respect to the ground state in the wavelength dependent resonator axes. We are, however optimistic that compact, steady - state, room temperature operated mid - infrared $E \rightarrow V$ - transfer- lasers, pumped by diode arrays can be built and used for applications like tracking of main constituents and pollutants of the atmosphere.

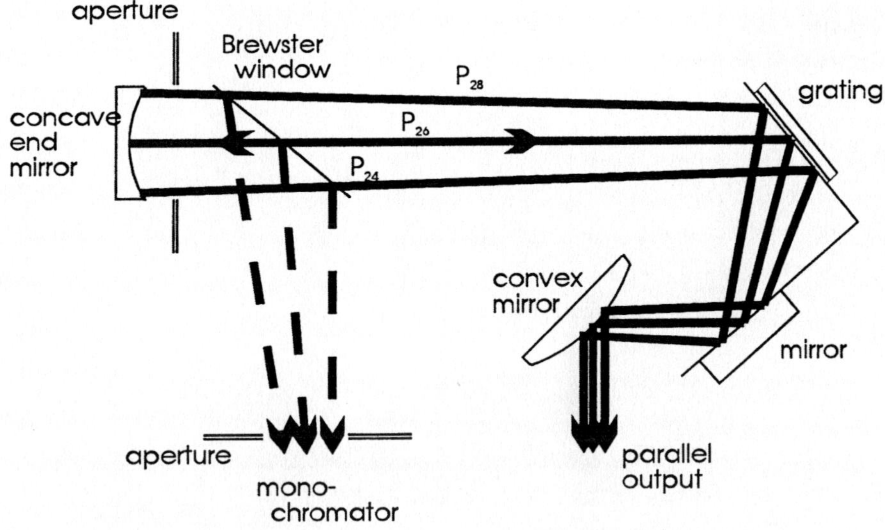

Figure 4: Observed modes arising from rotational transitions P_{24}, P_{26} and P_{28} in a cavity with wavelength - dependent resonator axes

References:

[1] H.K. Haugen, E. Weitz and S.R. Leone, J. Chem. Phys. **83**, 3402 (1985)

[2] C.R. Giuliano, L.D. Hess, J. Appl. Phys. **40**, 2428 (1968)

[3] L.E. Zapata, R.J. DeYoung, J. Appl. Phys. **54**, 1686 (1983)

[4] R.L. Pastel, G.D. Hager, H.C. Miller and S.R. Leone, Chem. Phys. Lett. **183**, 565 (1991)

[5] P.L. Houston, Advan. Chem. Phys. **47**, 381 (1981)

[6] A.B. Petersen, C. Wittig and S.R. Leone, Appl. Phys. Lett. 27, **305** (1975)

[7] A.B. Petersen, C. Wittig and S.R. Leone, J. Appl. Phys. **47**, 1051 (1976)

[8] A.B. Petersen, C. Wittig, J. Appl. Phys. **48**, 3665 (1977)

[9] H. Hofmann, S.R. Leone, J. Chem. Phys. **69**, 641 (1978)

[10] M.B. Faist, R.B. Bernstein, J. Chem. Phys. **64**, 2971 (1976)

[11] A.J. Sedlacek, R.E. Weston Jr., G.W. Flynn, J. Chem. Phys. **93**, 2812 (1990)

[12] J. Finzi, C.B. Moore, J. Chem. Phys. **63**, 2285 (1975)

[13] G. Guelachvili, J. Mol. Spectrosc. **79**, 72 (1980)

Intracavity CO-Laser Photoacoustic Trace Detection; CH$_4$ Production by Methanogenic Bacteria

F.G.C. Bijnen, F.J.M. Harren, J. Reuss
Department of Physics, University of Nijmegen
A.H.A.M. van Hoek, T.A. van Alen and J.H.P. Hackstein
Department of Microbiology, University of Nijmegen
Toernooiveld 1, 6525 ED Nijmegen, The Netherlands.

A liquid nitrogen cooled CO-laser in combination with an improved photoacoustic (PA) cell, placed intracavity, are employed to monitor low gas concentrations. As a first application, measurements on the CH$_4$ production of biogenic sources are reported.

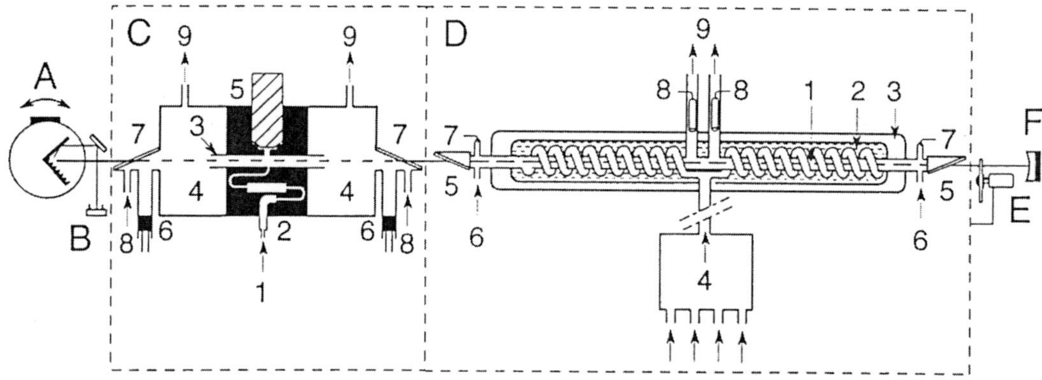

Figure 1: *Experimental set-up. The photoacoustic cell is placed intracavity in a CO-laser. The gas mixture enters the laser cavity through liquid N2 cooled spiralling tubes, from the left and right. A, Rotatable grating 230 l/mm; B, Powermeter; C, Photoacoustic cell; 1, Trace gas inlet; 2, 1/4 Lambda noise supressors; 3, Resonator; 4, Buffer; 5, Microphone; 6, Tunable air column; 7, Brewster window; 8, Buffer gas inlet; 9, Buffer gas – trace gas outlet; D, CO-laser; 1, Laser discharge tube; 2, Liquid N$_2$ reservoir; 3, Vacuum jacket; 4, Gas mixture barrel (CO, N$_2$, O$_2$ and He); 5, Brewster window; 6, He flow in; 7, Anode; 8, Cathode; 9, Exit laser gas to pump; E, Chopper; F, 100 % Mirror: R=10 m.*

Introduction

Our home-made liquid nitrogen cooled CO-laser [1] is tunable over 250 single line transitions between 4.8 μm ($v=3\rightarrow2$) and 7.9 μm ($v=35\rightarrow34$). The laser is employed in combination with an improved intracavity longitudinal PA-cell [2,3]. Using this set-up offers the possibility to measure a variety of gases at (sub) ppb level and to detect rapid changes in trace gas production (1 minute response). For (methane) CH_4 a detection limit of 1 ppb has been reached under practical conditions.

Atmospheric methane is an important component in the process of global warming and ozone depletion. The biogenic sources are responsible for the majority of CH_4 fluxes (>70%). The contribution of each different source is still a question of debate [4,5].

We have applied our system to detect CH_4 production by methanogenic bacteria (methanogens) residing in anaerobic freshwater sediments and in the hindgut of cockroaches, respectively.

Experimental set-up.

The experimental set-up is shown in figure 1. The laser consists of a $\varnothing$ 11.4 mm $\times$ 1.2 m pyrex discharge tube submerged in liquid nitrogen. The laser-gas mixture (CO, He, N_2 and O_2) flows longitudinally from the sides to the centre. The ZnSe brewster windows are continuously flushed by a He flow to avoid contamination. Cooling the laser to liquid nitrogen temperature (77 K) is advantageous for this results in higher laser power, a wide frequency range and single line emission [3,6] (see figure 2). Intracavity laser power can amount to 40 Watt at a wavelength of 5.2 μm. At the red side of the power spectrum most lasing transitions can still be operated at a power of 1 Watt. In extracavity operation the laser power is typically a factor of ten lower. We are able to operate the CO-laser for 5 days continuously. During this period intracavity power at e.g. 1304.97 cm^{-1} decreases by a factor of two. Probably this is due to solid CO_2 contamination on the inner wall of the laser discharge tube.

The photoacoustic effect is based upon energy conversion from light to sound energy. In our case, the modulated IR CO-laser radiation is being absorbed by a trace amount of e.g. CH_4 in air or nitrogen. Via relaxation the thus produced vibration energy is converted to sound. An asymmetric chopper modulates the radiation; a longitudinal PA-resonator amplifies the ensuing pressure rise due to the periodically produced heat passively and a microphone mounted at the anti-node of the resonator monitors the sound. The chopping frequency corresponds to an acoustic wavelength of twice the length of the resonator. At both ends the resonator is acoustically separated from the windows by buffers. The background signal emerging from the windows is further suppressed by means of an air column, in length tunable, at the window.

Absorption of (scattered) laser radiation can periodically heat up the wall of the PA resonator and forms therefore a source of background signal. By using a copper tube as resonator material, this absorbed heat is immediately dispersed by conduction. Absorption of radiation at the tube wall is minimized by a gold coating applied to the resonator.

Acoustic noise from the surrounding can also be a limiting factor for the sensitivity. Although the cell is acoustically shielded from the lab, noise still arrives at the microphone and limits the sensitivity of the system at this moment (see table 1).

The buffers are scaled to the dimensions of the resonator resulting in a total cell volume of 1.5 litre.

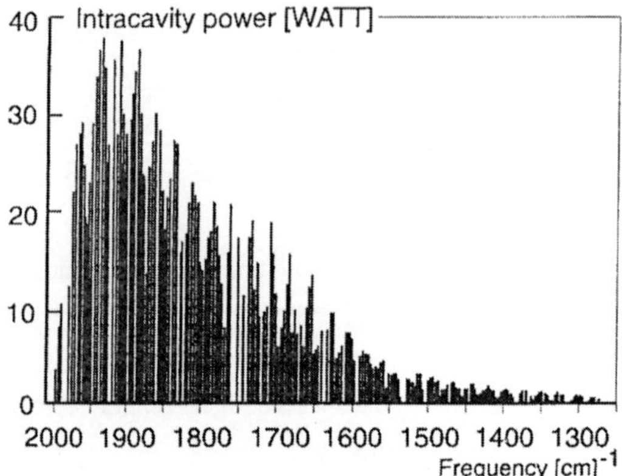

Figure 2:
Intracavity single line laser power of the CO-laser on 200 lines in the 5.0 to 7.8 μm wavelength range.

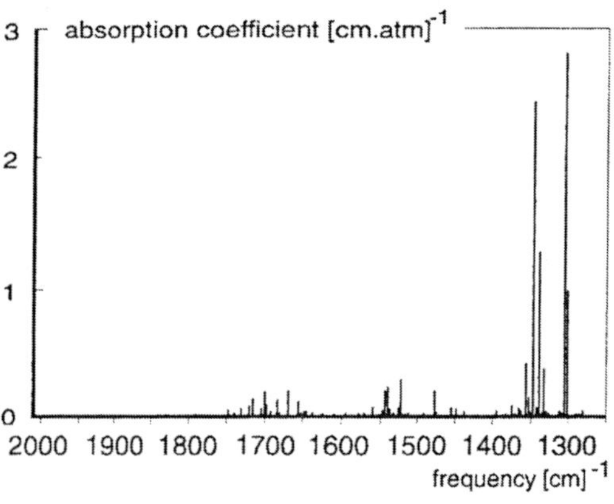

Figure 3:
Absorption coefficients of CH_4 in the CO-laser wavelength region. The peaks between $1400 \ cm^{-1}$ and $2000 \ cm^{-1}$ are due to water vapour.

Admitting the trace gas flow at the buffers (at a typical flow rate of 1.2 litre/hour) would result in a slow response of the system on changing concentrations. At the same flow rate, the trace gas inlet positioned at the centre of the resonator yields a response time of 1 minute. Noise entering the resonator at this inlet is effectively suppressed by 1/4 acoustic wavelength tubes in line with the trace gas flow.

Due to buffer gas diffusion back into the resonator the signal monitored by the microphone consists for 15 % of the trace gas present in the buffers. Therefore 85% of the signal is produced by the trace gas flowing from the centre of the resonator towards the buffers (at a flow of 1.2 l/hour). Since the residence time of the trace gas in the buffers is much longer than that for the resonator it causes a fairly constant background signal proportional to the measured average concentration. To reduce this effect two additional flows of 3 l/hour each, containing no trace gas, are admitted to the buffers. The trace gas concentration in the buffers is consequently diluted and the residence time is shortened by a factor of 5. Balancing the in and out flow of the buffers needs attention, but does not introduce severe problems. Choosing a flow system like described opens the possibility to cover the inner walls of the buffer with sound absorbing material without lengthening the response time of the cell and without creating problems due to trace gas sticking effects. Furthermore it offers the opportunity to put several acoustic resonators, in line with the laser beam, one behind the other, only separated by buffers and without extra Brewster windows.

A long term measurement consists of continuously switching between two (or more) laser lines, monitoring the signal and calculating the concentration(s). For the CH_4 measurements we used the ($v=33\rightarrow32$, $J=8\rightarrow9$) and ($v=33\rightarrow32$, $J=9\rightarrow10$) transitions of CO corresponding to 1308.01 cm^{-1} and 1304.97 cm^{-1} (see fig. 3). In this way a long term detection limit for CH_4 of 1 ppb has been reached corresponding to a minimal detectable biogenic production rate of 40 picomol per hour.

Approximate detection limits for some other biologically relevant gases have been determined. If no interfering gases are present we are able to measure CS_2 down to a limit of 10 ppt; acetaldehyde, SO_2, H_2O and NO_2 to 100 ppt; NO to 300 ppt; C_2H_2, C_2H_4, C_2H_6, N_2O, OCS and trimethylamine to 1 ppb and ethanol, pentane, DMS, methanethiol and NH_3 to 3 ppb. Our system is less sensitive for H_2S (100 ppb), CO and CO_2 (both 1 ppm). Water vapour can interfere with the detected trace gases. Before entering the PA-cell the gas flow is cooled down by means of a cold trap. In this way we are able to reduce the water vapour concentration to 1 ppm. In the case of measuring CH_4 this concentration does not present interference problems.

As tubing material we use PFA teflon to reduce memory effects in the case of trace gases like acetaldehyde and H_2S possessing a permanent dipole moment [7]. Indeed the response time for these gases on concentration changes was nearly the same as that for CH_4. Thus measuring these gases does not introduce extra problems due to sticking at the inner wall of the tubing material.

Results and Discussion of CH_4 measurements.

Natural wetlands and rice paddies may contribute up to 30% of the global emission of CH_4 [4,5]. Methanogens residing in sediments can be present in two forms; free-living or inside unicellular organisms (protozoa) [7,8]. Bromoethanosulfonic acid (BRES) is a chemical which inhibits the CH_4 biosynthesis. It is used to determine the unknown ratio of CH_4 production between free-living and endosymbiotic methanogens.

Table 1: *Performance of the PA-cell*

Resonator	$\varnothing$ 15 × 150 mm
Buffers	$\varnothing$ 100 × 100 mm
Resonance frequency	1000 Hz
Quality factor	40 ± 2
Cell constant	$(2.0 \pm .1) \cdot 10^3$ Pa cm/W (no flow)
	$(1.7 \pm .1) \cdot 10^3$ Pa cm/W (trace gas flow 1.2 l/hour, see text)
Sensitivity Microphone (B&K 1479)	1 V/Pa
Microphone noise	200 nV/$\sqrt{Hz}$
Acoustic background noise	2 μV/$\sqrt{Hz}$
Sensitivity of the PA-cell	$(1.0 \pm .05) \cdot 10^{-9}$ cm^{-1} (no flow)
(1 Watt laserpower)	$(1.15 \pm .05) \cdot 10^{-9}$ cm^{-1} (trace gas flow 1.2 l/hour)

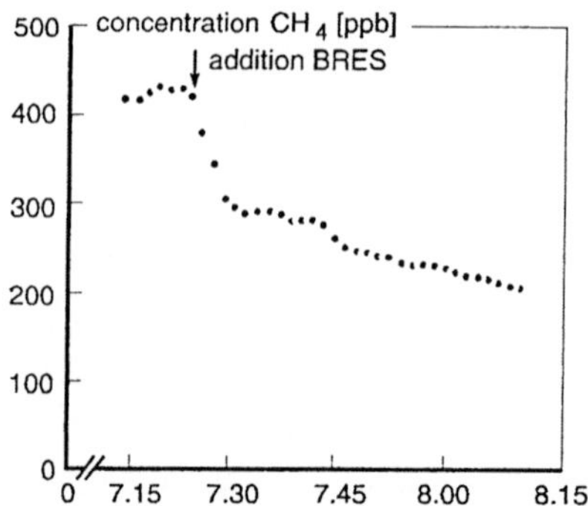

Figure 4:

Response of methanogens residing in 5 ml anaerobic freshwater sediment to the addition of 25 mM Bromoethanosulfonic acid (BRES).
Nitrogen flow 1.2 l/hour.

In figure 4 can be seen that the rather constant level of CH$_4$ production by methanogens changes after the addition of 25 mM BRES. The decrease of 30 % within 4 minutes, is probably caused by the immediate reaction of the free living methanogenic bacteria. Although biological interpretation of these measurements is still incomplete, the rapid response of our measurement system is demonstrated.

The contribution of methanogens living in the hindgut of insects like termites or cockroaches can amount to 20 % of the global emission of CH$_4$ [4,5]. Since nearly nothing is known about the regulation of the CH$_4$ synthesis in insects [10], we started to measure the CH$_4$ production by a single *Pycnoscelus surinamensis* (greenhouse roach) continuously over a period of 60 hours. Its response to a periodical light stimulus is shown in figure 5. The "light on" stimulus causes an increase of the CH$_4$ emission rate by about 40 ppb within one hour. After dissection of the cockroach we found a large number of flagellates with intracellular methanogenic bacteria in the hindgut. It is likely that the increase in the emission rate is caused by an enhanced biosynthesis of CH$_4$ in these type of protozoa. However it remains to be investigated, whether the protozoa responded directly to the stimulus or whether the cockroach controls the activity of the protozoa.

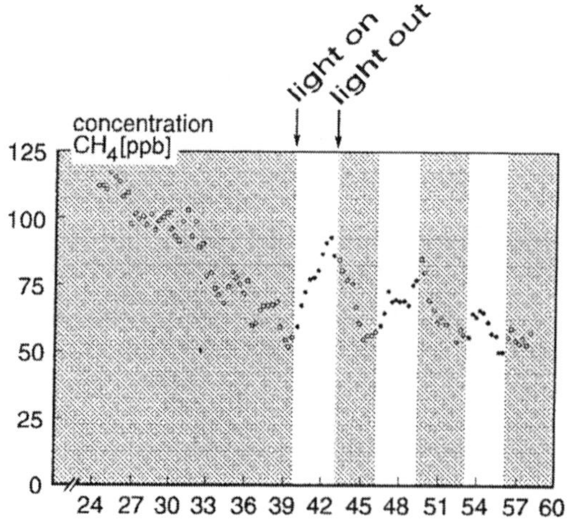

Figure 5:

Response of Pycnoscelus surinamensis (green roach) containing flagellates, to a modulated light source.
Airflow 1.2 l/hour.

References.

[1]   Development of a new CW Single-Line CO Laser on the $v'=1\rightarrow v''=0$ Band, B. Wu, T. George, M. Schneider, W. Urban and B. Nelles, Appl. Phys. B. **52** (1991) 163-167

[2]   Sensitive Intracavity Photoacoustic Measurements with a CO_2 Waveguide Laser, F.J.M. Harren, F.G.C. Bijnen, J. Reuss, L.A.C.J. Voesenek and C.W.P.M. Blom, Appl. Phys. B. **50** (1990) 137-144

[3]   A Liquid Nitrogen Cooled CO-laser in a Photoacoustic Set-Up Monitors Low Gas Concentrations, in "Photoacoustic and Phototermal Phenomena III", F.G.C. Bijnen, T. Brugman, F.J.M. Harren and J. Reuss, ed. D. Bicanic, Springer Series in Optical Science, **69** (Springer Verlag Heidelberg, 1992) 34-37

[4]   The Global Methane Budget, in "Microbial Production and Consumption of Greenhouse Gases: Methane, Nitrogen Oxides and Halomethanes", S.C. Tyler, Eds. J. E. Rogers and W. B. Whitman; (American Society for Microbiology, Washington DC, USA, 1991) 7-38

[5]   Contribution of Biological Processes to the Global Budget of CH_4 in the Atmosphere, in "Current Perspectives in Microbial Ecology", W. Seiler, Eds. M.J.Klug and C.A. Reddy; (American Society for Microbiology, Washington DC, USA, 1984) 468-477

[6]   CO-Laser Photoacoustic Spectroscopy of Gases and Vapours for Trace Gas Analysis, S. Bernegger and M.W. Sigrist, Infrared Phys. **30** (1990) 375-429

[7]   Cell Coatings to Minimize Sample (NH_3 and N_2H_4) Adsorption for Low-Level Photoacoustic Detection, S.M. Beck, Applied Optics **24** (1985) 1761-1763

[8]   Methane Production in Littoral Sediment of Lake Constance, B. Thebrath, F. Rothfuss, M.J. Whiticar and R. Conrad, FEMS Microbiology Ecology **102** (1993) 279-289

[9]   Abundance and Distribution of Anaerobic Protozoa and their Contribution to Methane Production in Lake Cadagno (Switzerland), S. Wagener, S. Schulz and K. Hanselmann, FEMS Microbiology Ecology **74** (1990) 39-48

[10]   Methanogenic Bacteria as Endosymbionts of the Ciliate *Nyctotherus ovalis* in the Cockroach Hindgut, H.J. Gijzen, C.A.M. Broers, M. Barughare and C.K. Stumm, Appl. Environ. Microbiol. **57** (1991) 1630-1634

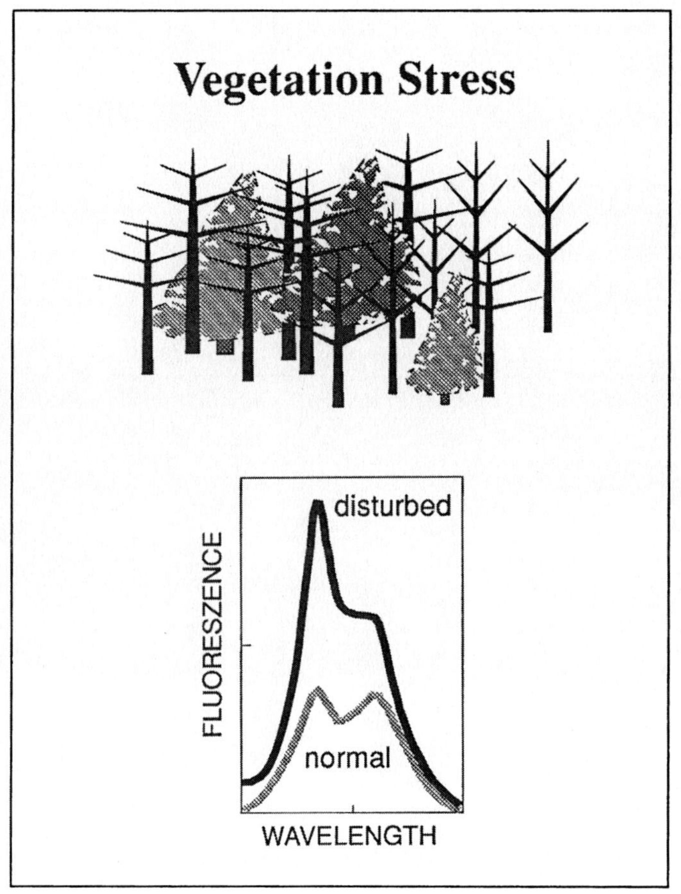

Vegetation Stress
disturbed
normal
FLUORESZENCE
WAVELENGTH

Lidar Remote Sensing of Vegetation Status:
The Link to Plant Physiology

G. Cecchi, L. Pantani, M. Bazzani;
I.R.O.E.-C.N.R., Via Panciatichi 64, I50127 Firenze, Italy.
V. Raimondi;
University of Florence, Electronic Engineering Dept., Via S.Marta 3, I50139 Firenze,
Italy.
P. Mazzinghi;
I.E.Q.-C.N.R., Via Panciatichi 56/30, I50127 Firenze, Italy.
R. Valentini;
University of Tuscia, Forest Resources Dept., Via S. Camillo De Lellis, I01100 Viterbo,
Italy.

INTRODUCTION - The remote sensing of laser induced fluorescence in terrestrial plants
has good potentials for species identification, green biomass estimation, leaf area index and
canopy structure (1), giving the actractive perspective for an airborne fluorescence lidar
operation.
Chlorophyll fluorescence has been used for many years as a powerful tool in plant
physiology to understand the primary events of photosynthesis and the stresses
development affecting photochemistry. The developed indexes are used to detect stresses
on intact leaves, both in laboratory and in natural conditions (2).
These physiological indexes derived by pulse modulated fluorimetry are not monitorable by
airborne remote sensing, requiring a dark pre-adaptation of leaves. Anyway, since
fluorescence spectral shape can be measured with remote sensing techniques, some
spectral parameters have been developed for the detection of physiological stresses (3). In
particular, the ratio of the fluorescence bands at 690 nm and 730 nm was considered as a
possible index of plant stresses (4), even if the physiological meaning of this spectral index
is still under debate.
This paper presents some results of laboratory and field measurements, obtained with the
near field systems (LEAF and PAM fluorometers - (5)) and the far field system (FLIDAR-
3 - (6)). The laboratory experiments point out some important factors affecting chlorophyll
fluorescence and, whenever possible, were compared with results of field and remote
sensing experiments.

EXPERIMENTAL - The fluorescence spectra show peaks at wavelengths that shift
slightly in the range of 5 nm for the F685 band and even 20 nm for the F730 band,
depending mainly on vegetation type and leaf structure. So, the term 'Red Fluorescence
Ratio' (RFR) is accepted to replace the well-known F690/F730 ratio.

Near field measurements of fluorescence signal were obtained with LEAF portable
fluorometer which measures the two bands at 685 ± 5 nm and 730 ± 5 nm. The first
experiment was performed on leaves of
Populus alba L. and *Quercus ilex* L.
enclosed in a controlled chamber and
exposed at different light levels. Diurnal
cycle measurements of fluorescence with
the LEAF fluorometer were collected on
trees of *Juglans regia* L. simultaneously to
the Photosynthetic Photon Flux Density
(PPFD). The data show a decrease of RFR,
increasing light, both for *Quercus ilex* and
Populus alba, and the same behaviour was
confirmed by a daily cycle of RFR taken in
the field on a leaf of *Juglans regia* L.
(fig.1). The fluorescence ratio decreases

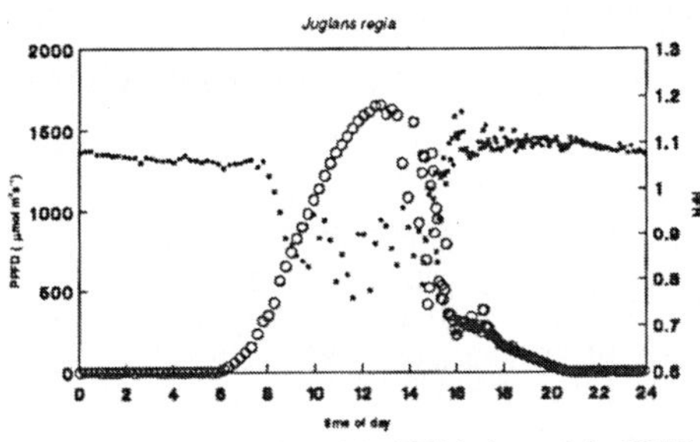

Fig.1 - Diurnal behavior of the RFR index and the PPFD.

during the day and reaches its minimum value at the highest value of PPFD. The decrease of RFR with increasing light level was found in all the experiments to be determined by a relative higher decrease of the 685 nm band with respect to the 720 nm one.

Furthermore, measurements were made simultaneously on a *Quercus pubescens* Wild. tree with the whole equipment to monitor the behaviour from normal to stress conditions. In particular, the stress was produced simply by cutting the tree branch, inducing mainly a water stress at first. The monitored leaves were enclosed in an environmental controlled cuvette for measurement of physiological parameters.

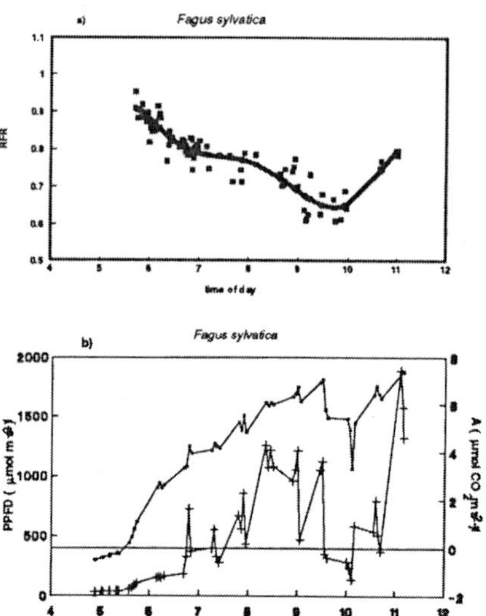

Fig.2 - Diurnal behavior of far field RFR measurements: a) RFR values during night-day transition; b) net photosynthesis and PPFD values of the target.

Remote sensing measurements were carried out on *Fagus sylvatica* L. and on *Quercus pubescens* Wild. trees with FLIDAR-3, detecting the complete spectrum from 500 nm to 800 nm at distances ranging between 15 and 60 meters and using a folding mirror settled just over the tree to simulate an airborne remote sensing experiment. The plants were monitored simultaneously with the near field equipment, including physiological measurements.

Fig.2a shows RFR values, obtained *in vivo* on a leaf of *Fagus sylvatica*. The fluorescence signal was monitored during several hours, observing the night-to-day transition. The relative PPFD values together with net photosynthesis measurements are shown in fig.2b. RFR shows an opposite behaviour compared to that of PPFD and PPFD variations, mainly due to meteorological conditions, induce variations on RFR. The near field measurements, obtained with LEAF fluorometer, show the same RFR behaviour confirming FLIDAR results.

Fluorescence spectra were also teledetected on unstressed and water stressed *Quercus pubescens* trees. The spectral analysis reveals (fig.3) that under stress conditions the peak at 690 nm is strongly decreased. Moreover, a small peak, placed by a gaussian fit at 713 nm, appears and remains in all the spectra obtained during the afternoon. This band is probably related to changes in pigment composition. Also in this experiment LEAF measurements confirmed the RFR behaviour detected by FLIDAR-3.

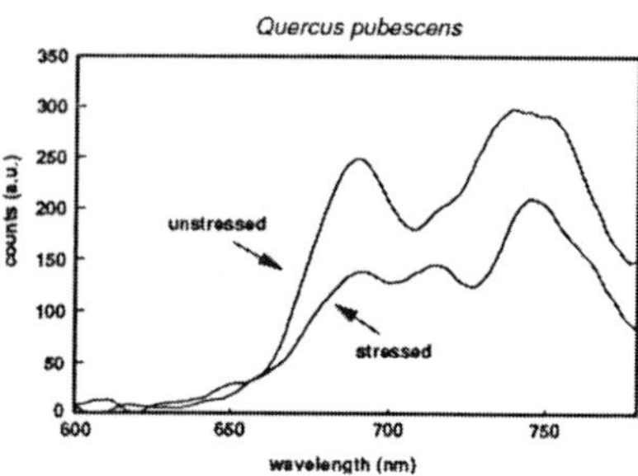

Fig.3 - Fluorescence spectra of *Quercus pubescens* under unstressed and water stressed conditions.

CONCLUSIONS - This work shows that chlorophyll fluorescence spectra are a sensitive tool for vegetation remote sensing. The connection between plant physiology parameters and RFR was demonstrated by several laboratory and field experiments. However, since many parameters affect RFR, the extraction of a vegetation stress index from remote sensing spectra is not easy. The first factor to be considered using the F690/F730 ratio is its relative dependence on the optical properties of leaves, especially the reabsorption of chlorophyll. This effect can be taken into account with a simple model using the combined information delivered by reflectance and transmission spectra (7). Furthermore, RFR was found dependent on environmental conditions, especially PPFD, showing a daily, cyclic variation that cannot be attributed to variations in chlorophyll concentration. This was also confirmed in laboratory experiments, with a direct relationship between RFR and light level.

Remote RFR measurements carried out in the far field with fluorescence lidar techniques were successful on different trees and different environmental conditions. The RFR variations showed the same behaviour of the laboratory experiments. The variations were also similar to the field measurements carried out on single leaves with LEAF fluorometer. The main difference between near field and far field measurements was found in the total fluorescence intensity variation with PPFD, during daily cycles. The FLIDAR measured a nearly constant fluorescence intensity, while the LEAF (and PAM) fluorometer showed the usual fluorescence quenching induced by the actinic light.

The fluorescence signal is generally quenched under light conditions by photochemical and non-photochemical quenching (8). Therefore lower fluorescence quantum yield is expected at increasing light levels.

A possible explanation of the observed contrasting behaviour is that the FLIDAR fluorescence signal is induced by an excitation pulse, whose duration is about 10 ns. This is much shorter than the pulse used by the LEAF fluorometer that is longer than 1 ms. Under dark conditions, when reaction centres are open, a short pulse is equivalent to a measurement of F_0 since most of excitation is trapped in the antenna chlorophylls. During high light condition most of the reaction centres are closed and the excitation gives a higher fluorescence signal, since the photochemical quenching has a longer time scale. It is not clear why the non-photochemical quenching seems not to affect laser-induced fluorescence signal under full sunlight conditions. Further research work is needed to understand the physiological processes that are at the basis of the observed discrepancies. In particular the different duration of the excitation pulses should be taken into account to analyse photochemical processes.

REFERENCES

(1) Chapelle, W.E., Wood, F.M., Mc Mutrey III, J.E., Newcomb, W.W. (1984), Laser-induced fluorescence of green plants. 1: A technique for the remote detection of plant stress and species differentiation, Appl. Opt. 23, 134-142.

(2) Demming, B., Winter, K. (1988) Characterization of three components of non photochemical quenching and their response to photoinhibition, Aust. J. Plant Physiol. 15, 163-177.

(3) Lichtenthaler, H.K. (ed) (1988) Application of Chlorophyll fluorescence, Dordrecht, Kluwer.

(4) Lichtenthaler, H.K., Rinderle, U. (1988) The role of chlorophyll fluorescence in the detection of stress conditions in plants, in CRC Critical Reviews in Analytical Chemistry, Vol 19 (suppl. I) S29-S85. CRC Press.

(5) Cecchi, G., Mazzinghi, P., Pantani, L., Valentini, R., Tirelli, D., De Angelis, P. (1992), Remote Sensing of chlorophyll *a* fluorescence on vegetation canopies: 1. near and far field measurements techniques, Int. J. of Rem. Sens. (in press).

(6) Cecchi, G., Pantani, L., Breschi, B., Tirelli, D., Valmori, G. (1992), FLIDAR: a multipurpose fluorosensor-spectrometer, EARSeL Advances in Remote Sensing 1, 72-78.

(7) Agati, G., Fusi, F., Mazzinghi, P., and Lipucci di Paola, M. (1992), A simple approach to evaluate the reabsorption effect of chlorophyll fluorescence in intact leaves, J. Photochem. Photobiol. (in press).

(8) Schreiber, U., Schliwa, U., Bilger, W. (1986) Continuous recording of photochemical and non-photochemical chlorophyll fluorescence quenching with a new type of modulation fluorometer, Photosynth. Res. 10, 51-62.

Time-Resolved Chlorophyll Fluorescence for Monitoring of Forest Decline

Herbert Schneckenburger, Werner Schmidt
Fachhochschule Aalen, Fachbereich Optoelektronik, Heinrich-Rieger-Str. 22/1, D-73430 Aalen, Germany

ABSTRACT

Aiming at clues of forest decline we studied prompt and delayed luminescence of spruce needles from the picosecond to the second time range using various custom made kinetic equipments. Both kinetics in the picosecond and in the seconds time range could be fitted by three exponentially decaying components yielding three amplitudes and three reaction constants each. Basically, all components showed a typical annual time course, independent of the degree of damage or air pollution. In addition, it turned out that on one hand the "slow" component of picosecond decay kinetics (decay time $\tau = 2.0\text{-}3.5$ ns) reflects some damage of the photosynthetic apparatus, which is mainly located in photosystem II. Similarly, in long term delayed luminescence in the seconds time range the "fast" component (decay time $\tau = 0.13$ s) obviously carries some information on the spruces' vitality. In general, healthy or declined spruces showed the highest photosynthetic efficiencies but also the most pronounced stress symptoms during the summer period - probalby due to high irradiance, drought and increased ozone concentrations.

1. INTRODUCTION

Trees are submitted to various stress factors, such as environmental pollutants, mineral deficiencies, climate or parasite infections, giving rise to the current phenomenon of forest decline. In our project for more than two seasons we studied various luminescence kinetics ranging from steady state measurements (*Kautsky curves*[1]) to the detection of decay kinetics after a single or after repetetive light pulses. In contrast to common biochemical procedures, luminescence measurements offer many advantages. They are fast, non-invasive and highly selective for specific plant pigments, and they require only small samples. Whereas picosecond kinetics monitor primary steps of energy transfer in both photosystems (**PS I, PS II**), long term delayed luminescence (*DL*) probes subsequent steps of the photosynthetic electron transport chain.

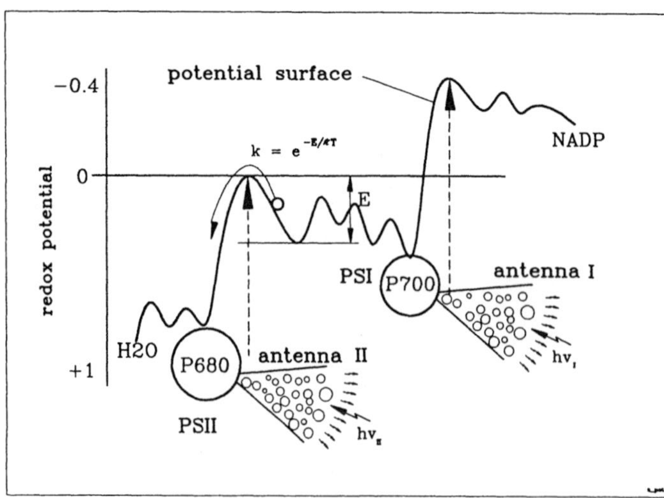

Fig.1 The generalized Z-scheme of the thylakoid membrane explaining prompt and delayed chlorophyll luminescence (antenna molecules: chlorophylls a and b, carotenoids; reaction centres P_{680}, P_{700}: photochemically active chlorophyll-protein complexes with main absorbancies at 680 and 700 nm). E is the activation energy, κ the Boltzmann factor and T the absolute temperature.

2 THEORY

Light is harvested by the antenna systems of the two photosystems **PS I** and **PS II** connected in series (**Fig.1**). The photons are funneled to the reaction centers which in turn drive the electrons from the donor side (H_2O) to the acceptor side (*NADP*) in a cooperative action across the thylakoid membrane. The trapping capability is lower than perfect giving rise to some reemission of prompt fluorescence. Therefore, picosecond spectroscopy was adopted to select the primary steps of energy transfer within the two photosystems[2]. After excitation by short laser pulses the fluorescence of the antenna chlorophyll molecules is expected to have a longer decay time, if the energy transfer to the reaction centers is obstructed.

The unexpected phenomenon of delayed light emission (*DL*) of photosynthecic organisms was discovered by Strehler and Arnold[3] in 1951. As compared to prompt fluorescence, DL turns out to be more complex and more susceptible to changes in experimental and particularly plant-physiological conditions. *DL* is indispensibly dependent on a functioning thylakoid membrane including electrical and pH-gradients, and is a concerted property of the photosystems. Mainly based on spectral evidence, the majority of researchers attribute *DL* exclusively to **PS II**, i.e. a reverse electron flow:

Absorption: ZChIQ (open state) + ε_F $\rightarrow$ Z⁺ChIQ⁻ (closed state)
Luminescence: Z⁺ChIQ⁻ (closed state) $\rightarrow$ ZChIQ (open state) + ε_F

with Z = electron donor, Chl = chlorophyll P_{680}, Q = chinoid electron acceptor and ε_F = exciton. Occasionally the *delay* is explained in terms of repetetive *retrapping* of excitons by the now open reaction center of *PS II*. However, there is a reasonable body of literature in favour of including also **PS I**, leaving this question unsettled, so far[4].

3. MATERIALS AND METHODS

3.1 Plant Materials

Needles were taken from 40-50 years old spruces (*picea abies*) at *Schöllkopf* near *Freudenstadt* (Black Forest) showing different degrees of both needle loss and yellowing (up tp 10 %: damage class 0; 10-25 %: damage class 1; 40-60 %: damage class 3). 12 spuces served for comparative measurements over a period of more than 2 years (Nov. 1989 - April 1992); another spruce (№ 33, damage class 2, together with a "healthy" reference) was used for light stress measurements starting in May 1993.

Small twigs containing the first and second age class of needles were harvested and kept humid till measurements, which were carried out 5 times per year. Only the results obtained from the second age class are reported in this article.

3.2 Methods

Fluorescence decay kinetics in the subnanosecond range were measured as described previously 2,5. Using a novel picosecond laser diode and a single photon counting equipment, chlorophyll emission was detected in the spectral ranges of 690-710 nm (**PS II**), 720-740 nm (**PS I** und **PS II**) and 695-800 nm. After deconvolution, a time resolution below 100 ps was obtained.

In addition, time-gated fluorescence spectra were measured in the ranges of 0-5 ns and 10-15 ns following the pulses of a dye laser (2 ns, 430 nm) pumped by the third harmonic of a Q-switched *Nd:YAG* laser. The spectra were recorded using an optical multichannel analyzer in combination with a gateable image intensifier (*Hamamatsu*, IMD, C4560) and a self-constructed microspectrometer (H. Schneckenburger, to be published). This equipment allowed fluorescent components of different decay times to be detected separately.

The luminescence decay of *DL* was measured in the range of about 20 seconds after single pulse excitation with red light at $\lambda \geq 665$ nm[6].

For the purpose of comparison, the chlorophyll concentrations of second year needles were determined according to Lichtenthaler and Wellburn[7]. These were in the range of 600-100 µg/ml fresh weight for damage class3, 800-1000 µg/g for damage class 1 and 1000-1700 µg/g for damage class 0, showing some seasonal variations (with maximal values in summer) and fluctuations between individual trees. The highest values were always found for spruce Nº 13 which appeared to be most "vital". All quantitative evaluations were carried out with visually intact green needles.

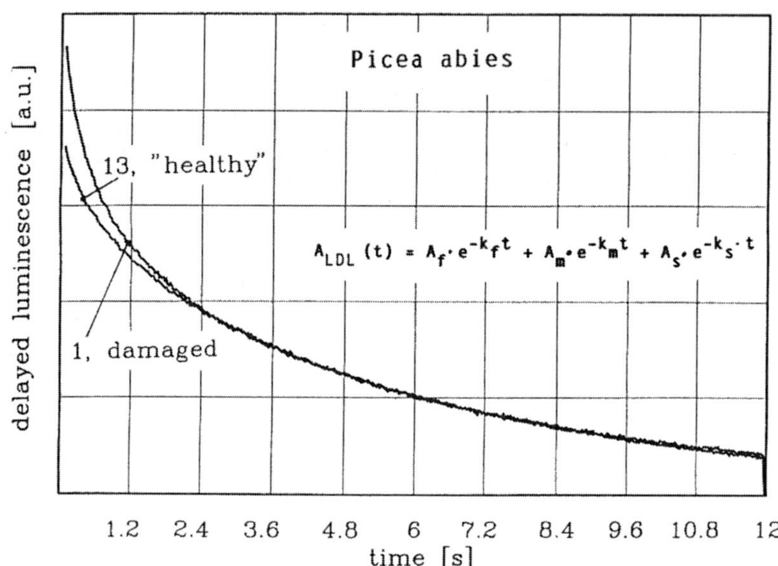

Fig.2 Decay kinetics of long term delayed luminescence as obtained from healthy (13) and declined (1) spruce needles.

4. RESULTS AND DISCUSSION

4.1 Delayed Luminescence

After single pulse excitation, *DL* of needles decays within some 10 seconds. **Figure 2** compares typical decay kinetics of small twigs of "healthy" (spruce Nº 13) and declined spruces (spruces Nº 1). These kinetics are perfectly fitted by a superposition of 3 exponentially decaying components, according to

$$A_{LDL}(t) = A_f\, e^{-k_f t} + A_m\, e^{-k_m t} + A_s e^{-k_s t}$$

with decay rates of k_f = 5-6 s⁻¹, k_m = 1.2 s⁻¹ and k_s = 0.15 s⁻¹ for a "fast", "middle" and "slow" component[6]. The relative amplitudes A_i of each component were found to depend on the individual tree and to vary more or less with the season (**Table 1**).

The most pronounced seasonal variation was found for the "fast" component A_f, as immediately verified in **Fig. 3**. In general, A_f showed the lowest values in early summer and the highest values in late autumn and winter. Parallel measurements of CO_2 consumption (B.Schweitzer, private communication) showed that A_f could be inversely correlated with the photosynthetic capacity. In general, declined spruces showed higher A_f-values with smaller annual variation than "healthy" trees: **Figure 3** compares spruces of damage class O (Nº 13), 1 (Nº 7) and 3 (Nº 1).

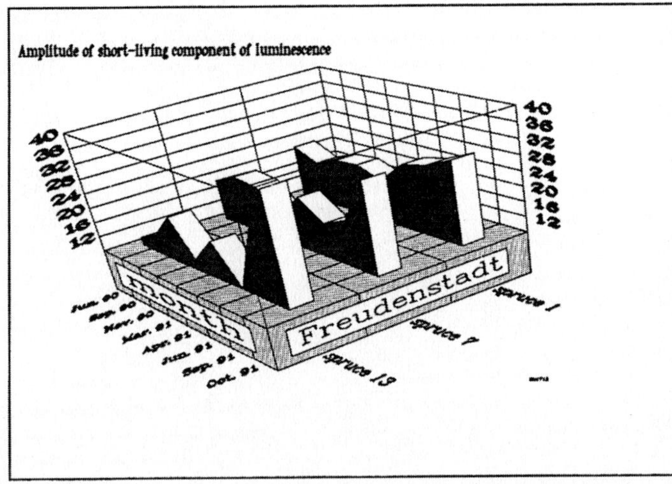

Fig.3 Time course of A_f [%] of long term delayed luminescence of green needles from a healthy (13), slightly (7) and a heavily damaged (1) spruce. Excitation at λ = 665 nm.

4.2 Prompt Fluorescence

Figure 4 shows the fluorescence decay curves of about 10 needles from a healthy spruce (lower curve) and a declined spruce (№ 33, damage class 2), as otained in the spectral range of 695-800 nm.

Obviously, the decay of the needles from the slightly damaged spruce occurs more slowly, even if they are green and healthy-looking. A more detailed analysis proved that all kinetics were composed by at least three exponentially decaying components with the time constants listed in **Table 2**.

TABLE 1

Decay times and amplitudes of *DL* for the various exponential components (τ=ln2/k)

	τ [s]	k [1/s]	amplitudes
τ_1 (kf)	0.13	5-6	A_f = 5-40%
τ_2 (km)	0.5	1.2	A_m = 20-30%
τ_3 (ks)	4.6	0.15	A_s = 40-60%

TABLE 2

Time constants of prompt chlorophyll fluorescence at different wavelengths.

wavelengths	τ_1	τ_2	τ_3
691-711 nm	~ 300 ps	~ 500 ps	2.0-3.5 ns
722-742 nm	~ 100 ps	~ 300 ps	3.5-5.0 ns
695-800 nm	100-200 ps	300-500 ps	2.0-3.5 ns

According to a comparison with literature values[8], the lifetime of about 100 ps may be attributed to photosystem I, that of 300-500 ps to the intact subunits of **PS II**, whereas the longer-lived component (2.0-3.5 ns) seems to arise from some "dead" chlorophyll or closed reaction centres. The relative intensity I3 of this long-lived component was evaluated for all descay curves according to

$$I_3 = A_3\, \tau_3\, /\, (A_1\tau_1\, +\, A_2\, \tau_2\, +\, A_3\, \tau_3)$$

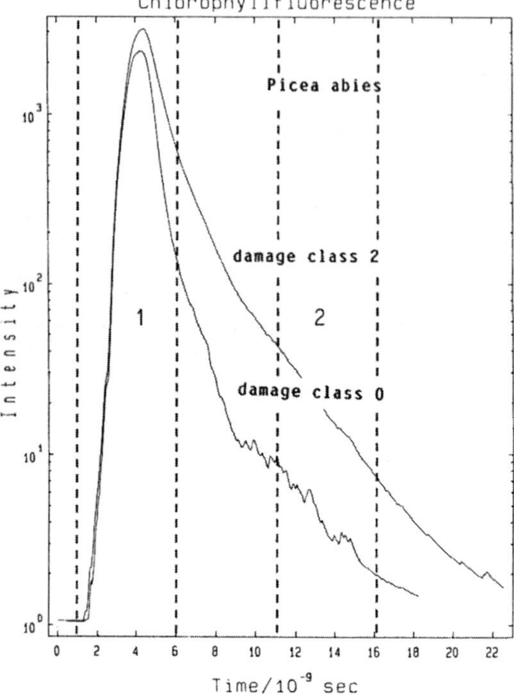

Fig.4 *Fluorescence decay of sunlight-exposed needles of a healthy reference spruce (lower curve) and a spruce of damage class 2 (N° 33) λ_{ex}: 680 nm, λ_{em}: 695-800 nm.*

where τ_i are the lifetimes and A_i the relative amplitudes obtained from tri-exponential curve fitting.

Preliminary measurements of needles of a damaged spruce (N° 33) under different light conditions showed considerably higher I_3 values for needles exposed to sunlight (up to 18%) than for those in shadow (about 2 %), thus indicating a certain light stress. **Figure 6** shows the time-gated emission spectra for 3 kinds of needles. In **Fig. 6A** (time gate 0-5 ns) the lower curve (spruce N° 33, light-exposed), middle curve (spruce N° 33, shadow) and upper curve (healthy reference spruce, shadow) roughly reflect the different chlorophyll concentrations. In contrast to this, **Fig. 6B** (time gate 10-15 ns) shows the fluorescence spectra of the component I_3 with a highest value of the light-exposed needles around 685 nm, the main emission peak of **PS II**. According to these spectra, the defect can be clearly localized in this photosystem.

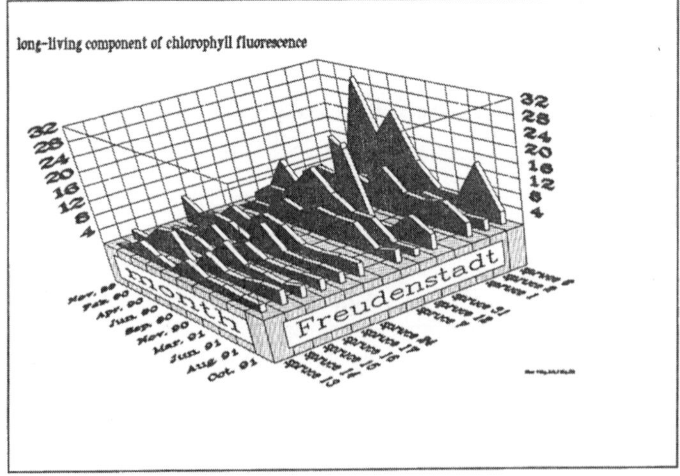

Fig.5 *Complete set of I_3-values [%] obtained from the spruces N° 13, 14, 15, 16, 17, 24: damage class 0; spruces N° 7, 12, 31: damage class 1; spruces N° 1, 2, 6: damage class 3.*

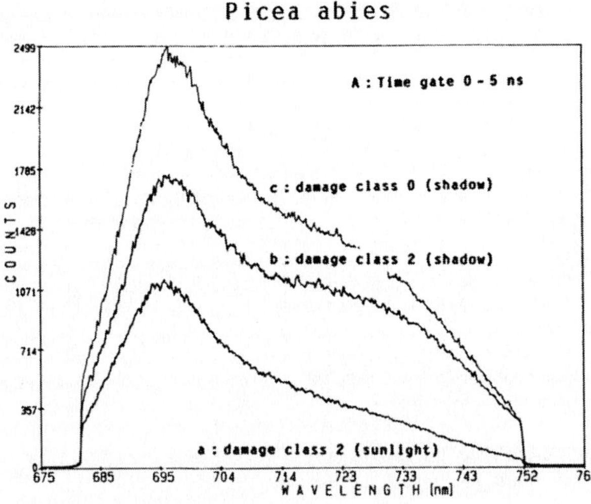

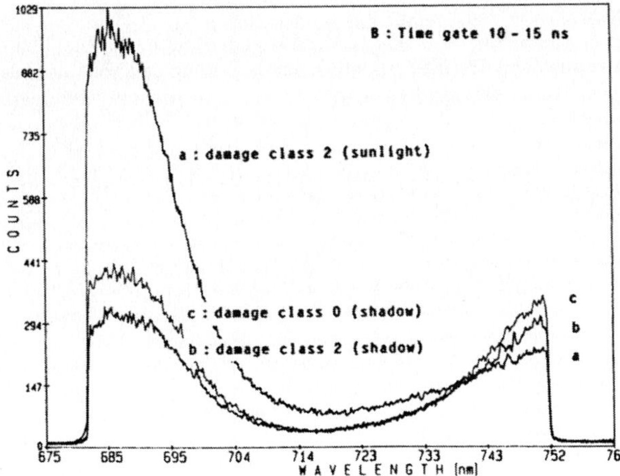

Fig.6 *Time gated fluorescence spectra of sunlight (a) and shadow (b) needles of spruce 33, as well as shadow needles of a healthy reference spruce (c); time windows for gated detection are indicated.*

Figure 5 presents a three dimensional survey of all I_3-intensities measured from Nov. 89 to Oct. 91. It clearly demonstrates the significance of the "slow" component of picosecond kinetics as a marker of the status of vitality of spruces. Trees are grouped according to their visually evaluated damage class in terms of yellowing and needle loos. Obviously, higher I_3-values were obtained for the spruces of damage class 1 and 3 as compared with those of damage class 0, thus confirming some reduced energy transfer within the photosystems of the damaged spruces.

A general (inverse) correlation between the chlorophyll concentration and the I3 values of the spruces in *Freudenstadt* was found. Both parameters seem therefore to be a measure for some defect within the photosystems. Some deviations from this general behaviour occur for slightly damaged spruces (e.g. № 31), where a high I_3 value indicates some reduction of energy transfer in photosynthesis, whereas the chlorophyll concentration is still high. Therefore, I_3 may be a very early indicator of tree damage.

Generally, photosynthetic defects (as reflected by I_3) seem to be more pronounced in summer than in winter (see **Fig.5**), which may be due to stress by high light doses, drought, and - possibly - increased ozone concentrations.

5. CONCLUSION

Two different kinds of luminescence parameters have been found: A_t as derived from *DL* and indicating the photosynthetic activity or vitality (thus avoiding laborious biochemical or gas exchange measurements) and I3 as deduced from prompt fluorescence and indicating a first stage of (invisible) stress or damage which is mainly located in **PS II**. Both sets of parameters may be useful tools in the fucture diagnosis of forest decline.

6. ACKNOWLEDGMENTS

This research was supported by the Projekt Europäisches Forschungszentrum für Maßnahmen zur Luftreinhaltung (PEF). The co-operation of T.Dienersberger, R.Hahn, K.-P.Irouschek, R.Hirsch and M.Bauer ist gratefully acknowledged.

7. REFERENCES

1. H.K.Lichtenthaler, C.Buschmann, U.Rinderle and G.Schmuck, "Applications of chlorophyll fluorescence in ecophysiology", Radiat.Environ.Biophys., Vol. 25, pp. 297-308, 1986.

2. H.Schneckenburger and W.Schmidt, "Time-resolving luminescence techniques for possible detection of forest decline (II) Picosecond chlorophyll fluorescence", Radiat.Environ.Biophys., Vol.31, pp. 73-81, 1992.

3. B.L.Strehler and W.Arnold, "Light production by green plants", J.Gen.Physiol. Vol.34, pp. 809-820, 1951.

4. W.Schmidt and H.Senger, "Long-term delayed luminescence in Scenedesmus obliquus". I. Spectral and kinetic properties. Biochim. Biophys. Acta, Vol.890, pp. 15-22, 1987.

5. H.Schneckenburger, W.Strauß, A.Rück, H.K.Seidlitz and J.M.Wessels, "Microscopic fluorescence spectroscopy and diagnosis", Opt. Eng., Vol. 31, pp. 995-999, 1992.

6. W.Schmidt and H.Schneckenburger, "Time-resolving luminescence techniques for possible detection of forest decline (I) Long term delayed luminescence", Radiat. Environ. Biophys., Vol. 31, pp. 63-72, 1992.

7. H.K.Lichtenthaler and A.R.Wellburn, "Determination of total carotenoids and chlorophyll a and b of leaf extracts in different solvents", Biochem.Soc.Transact., Vol.603, pp. 591-592, 1983.

8. A.R.Holzwarth,"Time-resolved chlorophyll fluorescence - what kind of information does it provide? in: Applications of Chlorophyll Fluororescence (H.K.Lichtenthaler,ed.), Kluwer, Dordrecht, pp. 21-31, 1988.

Technical Aspects of an Airborne Vegetation Fluorescence Lidar System

H.G. Dahn, W. Lüdeker, K.P. Günther
DLR Institute for Optoelektronik, PO Box 1116
82230 Weßling, Germany

This paper presents a general overview of the technical requirements for a vegetation fluorescence lidar and discusses the basic equations for estimating signal strength and signal to background ratio for a coaxial system.

Remote detection of laser induced plant fluorescence will be an important operational tool for modern agriculture and general observation of vegetation. When operated from an airplane large vegetation areas can be covered in a short time and their physiological state can be monitored.

Figure 1 shows the airborne situation and a typical vegetation fluorescence spectrum of maize. The main features of a plant fluorescence spectrum are the two bands in the red at 730 nm and 685 nm which are due to the red fluorescence of chlorophyll $\underline{a}$ and the broad blue fluorescence band centered at 440 nm (with 355 nm excitation). The pigments that emitt the blue fluorescence are not yet clear. Most promising candidates are chlorogenic acids, caffeic acid and similar phenolic substances for the blue fluorescence, whereas the green fluorescence peak at 530nm might be emitted by the alkaloid berberine. [Lichtenthaler et. al 1991]. The simple nature of the spectrum suggests that a discrete measurement approach might be sufficient. The main advantage of a discrete setup is the lower data rate and the possibility of on-line calculation of the data.

For airborne operation the coverage along the line of flight should be complete. At a typical flight speed of 300 km/h in about 300 m height and a desirable spot size of 80 cm on ground the repetition rate of the laser system has to be at least 100 Hz to achieve this requirement. The emitted fluorescence intensity of the plants is very low with respect to the inciding laser energy. The vegetation lidar equation gives the basic relation for the intensity at the receiver:

$$dF(\lambda,\Delta\lambda) = N/(\pi\, r_L^2) \cdot \alpha(c,\lambda_L)\, \eta\, \Phi\, (1 - \rho(\lambda,c))\, \tau \cdot \Omega_T/4\pi \cdot da \qquad \text{(Eq. 1)}$$

with the following symbols:

N	:	number of laser photons / pulse
ω_L	:	divergence of laser beam [rad]
$\alpha\,(c,\lambda_L) = 0.8$	:	absorptance of vegetation at λ_L
$\eta\,(N) = 0.005$	:	fluorescence efficiency
$\rho(\lambda,c)$	:	reabsorption factor
c	:	chlorophyll concentration
$\Omega_T = \quad A_T / H^2$	:	solid angle of telescope [sr]
H	:	flight height above zero [m]
A_T	:	area of telescope [m^2]
ω_T	:	divergence of telescope [rad]
$dF(\lambda, \Delta\lambda)$	:	fluorescence signal per unit area
da	:	infinitesimal area on ground, observed by the telescope
τ	:	transmission of optics
$\Phi(\lambda,\Delta\lambda)$	:	spectral weight function

The equation consists of three groups. The first part is the laser-photon density on ground. The second part of the equation is slightly different from the well known case of algae fluorescence in water [Measures 1984].

The percentage $\alpha(c,\lambda_L)$ of photons absorbed by the vegetation is multiplied by the fluorescence quantum yield $\eta(N)$, which is the percentage of the absorbed photons converted to fluorescence photons. As the vegetation can be regarded mainly as a hard target with a high chlorophyll concentration located in a relatively small area (chloroplasts). A photon that is absorbed and produces fluorescence is either reabsorbed (and has not to be taken into account furthermore) or emitted. As the total fluorescence quantum yield η is the integral $\int\Phi(\lambda)\,d\lambda$ over all wavelengths, the probability of detecting fluorescence photons at a distinct wavelength λ with bandwidth $\Delta\lambda$ is given by the spectral weight function $\Phi(\lambda,\Delta\lambda)$.

The term $\rho(\lambda,c)$ describes the percentage of reabsorbed fluorescence photons at wavelength λ. The reabsorption depends on the chlorophyll concentration and the wavelength. At 730nm the reabsorption factor is nearly zero, but at 690nm reabsorption can be clearly observed in the spectrum. [Dahn et. al 1992] τ is the transmittance of the receiver optics.

The third part of the equation regards the solid angle of the telescope and as is described:

$$r_L{}^2 = d^2/4 = (\omega_L\,H)^2/4$$

which yields to the equation:

$$dF(\lambda,\Delta\lambda)=N/(\,\omega_L\,H\,\pi\,)^2 \cdot \alpha(c,\lambda_L)\,\eta\,\Phi\,(1 - \rho(\lambda,c)) \cdot \Omega_T\,\tau\,da \qquad (Eq.1a)$$

The integration is carried out over the area observed by the telescope. If the divergence of the telescope is equal or greater than the divergence of the laser beam, the integration results in:

$$F(\lambda,\Delta\lambda) = N\,\,\alpha(c,\lambda_L)\,\eta\,\Phi\,(1 - \rho(\lambda,c))\,\Omega_T\,\tau\,/\,(4\pi) \qquad (Eq.\,2)$$

If the divergence of the telescope is less than the laser beam divergence, it yields:

$$F(\lambda,\Delta\lambda)_{T/L} = N\,\,\alpha(c,\lambda_L)\,\eta\,\Phi\,(1 - \rho(\lambda,c))\,\Omega_T\,\tau\,/\,(4\pi)\,\omega_T{}^2/\,\omega_L{}^2$$

which is basically (Eq. 1a) multiplied by the squared ratio of telescope divergence to laser divergence.

$$F(\lambda,\Delta\lambda)_{T/L} = F(\lambda,\Delta\lambda) \cdot \omega_T{}^2/\,\omega_L{}^2 \qquad (Eq.\,3)$$

The detected signal $S(\lambda,\Delta\lambda)$ is always a superposition of the fluorescence signal, the solar radiation reflected by the vegetation and the electronical noise of the opto-electronical device.

$$S(\lambda,\Delta\lambda) = F(\lambda,\Delta\lambda) + R(\lambda,\Delta\lambda) + N \qquad (Eq.\,4)$$

$R(\lambda,\Delta\lambda)$:   number of reflected photons of global irradiation
N :   electronical noise

Electronic noise should be neglected for a first estimation of signal and background. The solar radiation reflected by the vegetation and received by the detector can be described by the following formula:

$$R(\lambda,\Delta\lambda) = G(\lambda)\,A_T/2\,\pi\,H^2 \cdot \pi\,(\omega_T\,H)^2/4 \cdot \xi(\lambda)\,\tau \cdot \Delta\lambda\,\Delta t$$

and thus:

$$R(\lambda,\Delta\lambda) = G(\lambda)\,A_T\,\omega_T{}^2\,\xi(\lambda)\,\tau\,\Delta\lambda\,\Delta t\,/\,8 \qquad (Eq.\,5)$$

$G(\lambda)$:   number of global irradiance photons at wavelength λ per unit area and unit time (G(685nm,10nm) = 5.2 $\cdot$ 10^{19} [photons /(m^2 s)] (air mass 1),G(730nm,10nm) = 4.8 $\cdot$ 10^{19} [photons /(m^2 s)] (air mass 1)
$\xi(\lambda)$:   reflection coefficient (ξ(685nm) = 0.1, ξ(730nm) = 0.5), $\Delta\lambda$:bandwidth of detector, Δt:gate width of the detector

For detection of algae fluorescence, this part of the signal can be neglected, because water is almost dark at all visible wavelenghts. But for vegetation the background signal is the major limiting factor for a vegetation fluorescence lidar. In the frame of phase I of the EUREKA Project Lasfleur (EU380) two systems were developed by DLR. DLidaR-1 is a system based on an intensified diode array as detector unit. It was constructed for near field measurements. In contrast DLidaR-2 is a ground based far field fluorescence lidar system based on gated photomultipliers. It can operate up to a maximum distance of 300m, whereas most of the time it is used over a range of about 90m. For the future a flight system is planned on base of DLidaR-2, which is called in this text DLidaR-F. Its flight height shall be 300m. [Günther et al. 1991]

	DLidaR-2	DLidar-1	DLidaR-F
λ_L [nm]	355	355	396
E_I [mJ]	35	35	15
ω_L [mrad]	10	10	2,67
ω_T [mrad]	10	175	2,67
r_T [cm]	10	0,05	20
Ω_T [μsr] @ 40m	19,6	5×10^{-4}	78,5
G[photon x 10^{19}/m^{2}s]	4,8	4,8	4,8
N[photon x 10^{16}]	6	6	10
gatetime [ns]	50	32	50
F(730) [photon x 10^5]	374,3	$9,5 \times 10^{-3}$	2500
R(730) [photon x 10^4]	47	0,22	13,74
S / B	80	0,43	1819

Table 1. : Signal to background for the current configurations of DLidaR1 and DLidaR2.
Detection bandwidth $\Delta\lambda$ = 10 nm, spectral weight function $\Phi(730)$ = 0.1, absorption α (c,355nm) = 0.80, fluorescence quantum yield η (N) = 0.005, transmission in the receiver optics τ = 1.0 and the reabsorption factor ρ(730nm,c) is set 0.

As the DLidaR1 was developed for near field investigations up to 10m, it is not surprising that the signal to background ratio is no more acceptable at 40 m. The reason for that is the small entrance of its fiber optic.
The DLidaR2 is able to operate at maximum distances up to 300 meters. With a divergence of 10mrads, it delivers an excellent performance up to 90 meters, which is an adequate distance for ground based operations. The signal to background operation is 12 for day operation. By narrowing the laser spot the maximum distance can be increased furthermore.
Theoretically the signal to background ratio could be changed by increasing the laser pulse energy. The main limit in increasing the photon density at the target is limited by the annihilation criterion. Annihilation means, that the photosynthetic apparatus is oversaturated by photons which yields to a decrease of the fluorescence quantum yield in a non-linear manner. A spectrum taken in this range surely does not reflect the physiological state of the plant. In [Campillo et. al. 1976] this criterion is quantified as $N/cm^2 < 3 \cdot 10^{13}$ [photons/cm^2] per pulse in 100ps. If transferred to our DlidaR systems one can calculate, if a pulse length of 1ns is taken, that the maximum energy at 355nm wavelength is 1J, which is far beyond all proposed intensities. Thus the variation of the telescope divergence is not a critical factor with the proposed laser energy for the two systems.
For both systems, DLidaR-2 and the proposed DLidaR-F,(parameters shown in table1) one can calculate from (Eq.2 and Eq.5) the decrease of the signal to background ratio with increasing distance (Fig. 2). Regarding a S/B ratio of 10 as sufficient, the maximum measuring distance is 90m for DLidaR-2.
The divergence directly affects the signal to background ratio which can be seen from (Fig.3). The signal to background ratio is plotted against distance, parametrized by different divergences. All other data are given in table 1. Assuming the same S/B ratio and a laser divergence of 2.7 mrad one can see that the DLidaR-2 can be operated up to a distance of 500m. It should be mentioned, that the telescope divergence does not influence the S/B ratio.
For evaluating the quality of the signal we have to calculate the absolute number of fluorescence photons (Fig. 4) detected at the proposed flight height of 300m. For DLidaR-F the received intensity is always higher than for DLidaR-2, due to decreased laser divergence. The maximum laser energy needed is just 15 mJ at DLidaR-F, with a repetition rate of 100 Hz, in order to cover the whole area along the flight line. As better laser sources will become available in the near future, the signal (at only 50 mJ

laser energy) will even rise, and the maximum distance to the target will become 500m. As one can see, the feasibility of an operational flight system is only a matter of the technological development and can be achieved by reasonable effort. The proposed laser wavelength is about 397nm for optimal excitation of blue and red fluorescence. The available laser sources are tripled Nd:YAG with Raman shifter, dye lasers or doubled Ti:Saphire Lasers.

This work was financially supported by the German Ministery of Research & Technology (BMFT) in the frame of the EUREKA project LASFLEUR (EU380) under the contract number 0339290A.

Literature:

[Campillo et. al. 1976]   Campillo, A.J., Sharpiro, S.L., Kollmann, V.H. and Winn, K.R. (1976), Biophys. J. 16, 93-97

[Measures 1984]   Measures, R.M., Laser Remote Sensing, 1984, Wiley & Sons., New York, USA

[Lichtenthaler et al.1992] Lichtenthaler, H.K., Lang, M., Stober, F., Proc. of the 5th Int.Coll. - Physical Measurements and Signatures in Remote Sensing, Courchevel, France (ESA SP-319, May 1991) p. 727ff.

[Dahn et. al 1992]   Dahn, H.G., Günther, K.P., Lüdeker, W., EARSeL Advances in Remote Sensing, Vol. 1, No. 2 - II, 1992, 12-19

[Günther et al 1991]   Günther, K.P, Lüdeker, W., Dahn, H.G., Proc. of the 5th Int.Coll. - Physical Measurements and Signatures in Remote Sensing, Courchevel, France (ESA SP-319, May 1991) p. 723ff.

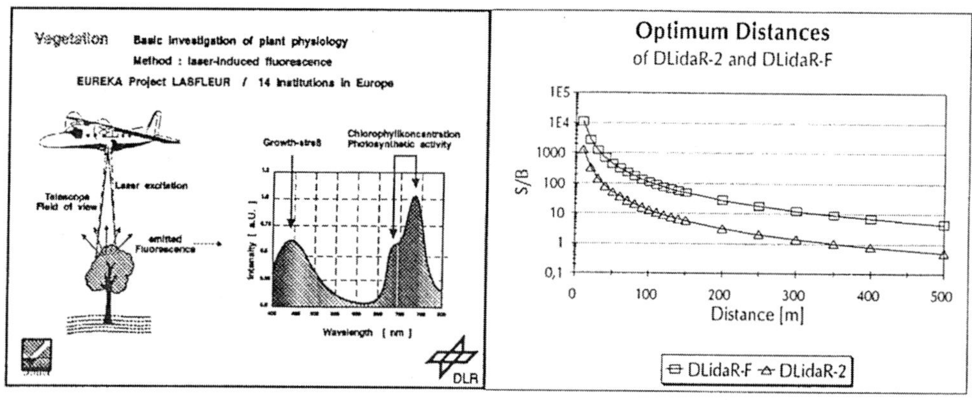

Fig. 1 Situation and objectives of an airborne vegetation fluorescence lidar.

Fig. 2 S/B-ratio for DLidaR-2 and DLidaR-F

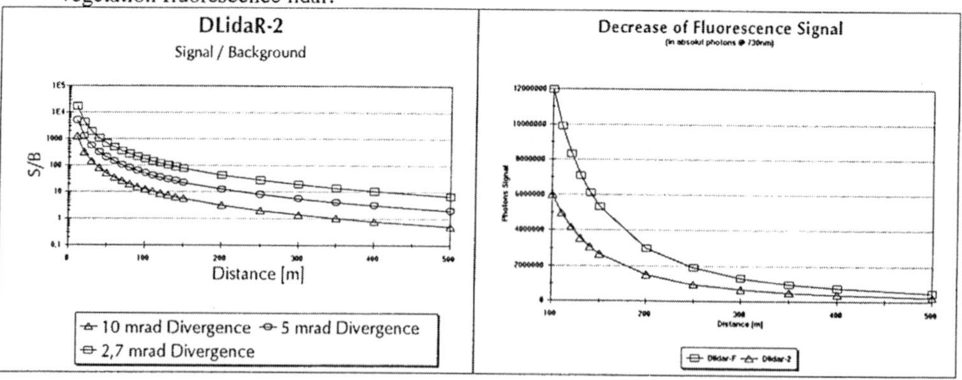

Fig. 3 S/B Ratio for DLidaR-2 with different laser divergency.

Fig. 4 Decrease of the fluorescence signal with distance.

Picosecond Fluorescence Decay and Backscattering Measurements of Vegetation over Distances

Y. Goulas[1], L. Camenen[2], G. Schmuck[3], G. Guyot[2], F. Morales[1] and I. Moya[1].
[1]*Laboratoire d'Utilisation du Rayonnement Electromagnétique, Orsay, France.*
[2]*Institut National de la Recherche Agronomique, Montfavet, France.*
[3]*Institut for Remote Sensing Applications, JRC, Ispra, Italy.*

Plant activity and growth are affected in a complex manner by changes in their environmental conditions such as those of light, water availability, temperature, CO_2 or the presence of some kinds of pollutants. An understanding of these interactions at a large scale (field, land or global scale) can only be made if remote sensing techniques are used. In the last decades, several spectroscopic methods have been considered for characterizing plant status or assessing biomass productivity. Among these methods the use of fluorescence signal emitted by plants appears as a promising tool.

Origin of fluorescence
This fluorescence emission occurs after the absorption of light by the photosynthetic pigments (carotenoids, chlorophyll a and b) (for a review see [1]). The absorbed energy is rapidly transferred to chlorophyll a and is used to drive photochemical reactions that provide the redox potential required for electrons transfer from water to CO_2. This transfer is produced by two photosystems working serially. Under optimum situations most of energy quanta absorbed by photosystem II are converted into photochemical energy [2]. But, under most of natural conditions, a more reduced part of the absorbed quanta is converted into photochemistry, and the remainder part is mainly dissipated as heat [3]. However, a small part of the energy is also re-emitted as fluorescence. It has been shown that the quantum yield of fluorescence, i.e. the number of light quanta emitted as fluorescence divided by the number of quanta absorbed by the leaf, is affected by the efficiency of the photochemical conversion and heat dissipation [4, 5] and thus can serve as a plant status indicator.
Fluorescence of leaves under natural conditions can be measured at distances using excitation with a short laser pulse and detection with a fast photomultiplier [6]. But remote determination of the quantum yield of fluorescence is very difficult, because the intensity of the detected fluorescence can be affected by poorly defined parameters. Among them we can find: leaf surface reached by the laser beam, amount of re-absorbed fluorescence, which depends on chlorophyll concentration in leaves, canopy structure, which affects light propagation, and distance of operation.
To overcome these difficulties, it has been proposed to measure the average fluorescence lifetime instead of the intensity for quantum yield determination [7]. Indeed, it has been shown that the average lifetime stays almost proportional to the quantum yield of fluorescence under most experimental conditions [8, 9]. The fluorescence lifetime, which represents the mean duration of the fluorescence emission after an ultra-short pulse, is not affected by varying experimental conditions such as leaf surface or chlorophyll concentration. It seems to be a well-suited parameter to access plant status at distances.
We describe here the newly developed picosecond LIDAR (LIgth Detection And Ranging) system of the LURE laboratory designed for the remote measurement of the fluorescence lifetime of plant canopy. We show that fluorescence lifetime can be used as a stress indicator and that its determination is possible even if the canopy structure is complex.

Instrument description.
Fig 1 presents a schematic representation of the mobile LIDAR system. The principle of the instrument is based on the excitation of a small portion of the canopy with a short laser pulse (35 ps duration - 1 mJ at 355 nm) provided by a tripled YAG laser. The fluorescence emission is detected

with a high speed crossed field photomultiplier (Sylvania Model 502). The laser beam is pointed towards the top of the canopy by a flat mirror. Ligth emitted from the canopy is collected trough a 0.4 m diameter Fresnel lens. The fluorescence signal is selected with a red highpass filter (>665 nm) and the backscattered signal with an inteference filter at 355 nm. The current pulse at the output of the photomultiplier is digitalized with a high bandwidth transient analyzer (Tektronix SCD1000 - 1Ghz Bandwith).

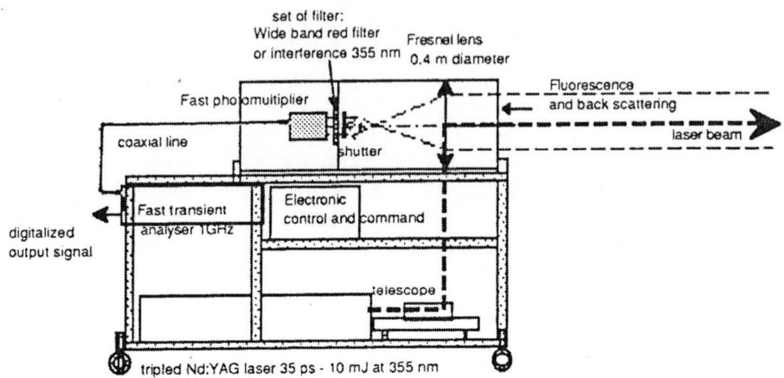

<u>Fig 1</u>: Mobile picosecond LIDAR system of the LURE laboratory for vegetation monitoring.

Principle of measurement.
Fig 2 shows the fluorescence emission of a single sugar beet leaf as a function of time after excitation with the laser pulse. It also shows the instrumental response recorded by looking at the backreflected light at the same wavelength as the excitation pulse. Experimental conditions were designed to simulate a detection from 50 m (actual distance of detection 4.5 m). One can see that the fluorescence signal is delayed compared to the backscattered signal. The delay value, or the fluorescence lifetime, can be computed if we consider that the measured fluorescence decay Fe(t) is the convolution product of the instrumental function I(t) with the actual fluorescence decay F(t):

$$F_e(t) = F(t) \otimes I(t)$$

F(t) is modelized as a sum of exponential decays

$$F(t) = \sum A_i \exp(-\frac{t}{\tau_i})$$

Deconvolution of the former expression, or the determination of parameters Ai and τi, is performed with a least squares method using the Marquardt search algorithm for non-linear parameters [10, 11]. Fig 2 shows the difference between the experimental decay Fe and the recalculated decay Fc. With the indicated parameters, the residue function is close to zero and the average lifetime of fluorescence can be computed as:

$$\tau = \frac{\sum A_i \tau_i^2}{\sum A_i \tau_i} = 0.34 \text{ ns}$$

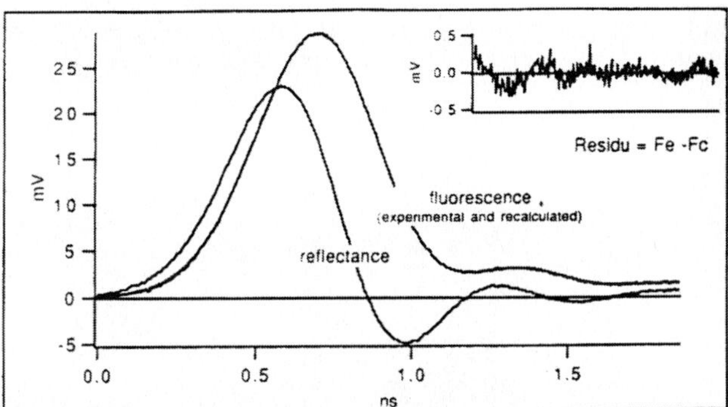

Fig 2: Fluorescence and backscattered ligth of a single sugar beet leaf after a picosecond laser pulse. The experimental decay Fe(t) is best modelized if F(t) is of the form:

$$F(t) = a_1 \exp(\frac{-t}{\tau_1}) + a_2 \exp(\frac{-t}{\tau_2})$$

with a_1=85, a_2=15, τ_1=0.168 ns and τ_2=0.60 ns. With these parameters, the recalculated fluorescence decay is almost identical to the experimental decay.

Effect of stress on fluorescence lifetime

When sugar beet grows in an iron deficient medium a stress develops which largely affects photosynthesis [12]. Fig 3 shows that this kind of stress can be detected by monitoring the average fluorescence lifetime. With the development of the stress, chlorophyll concentration becomes lower and the average lifetime increases significantly.

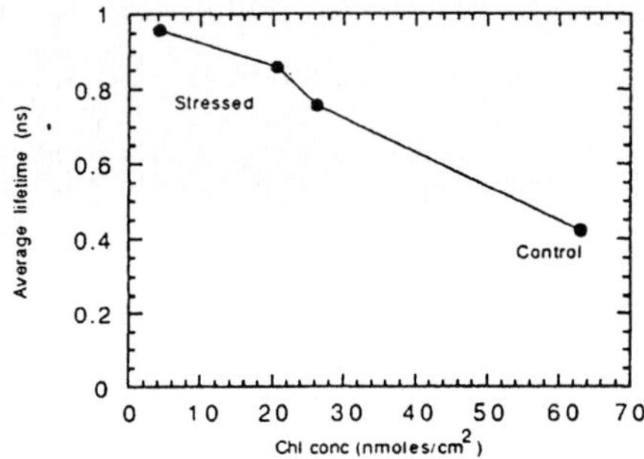

Fig 3: Variations of the average lifetime of fluorescence of sugar beet leaves in iron deficient crops. Measurements have been performed under an actinic illumination of 270 μmoles photons/m^2/s, which corresponds to natural conditions under moderate ligth.

Mesurement on a complex canopy

When performing measurements on a whole plant canopy, the laser spot is unlikely to reach a single plane leaf but several leaves separated one from the other by a distance of a few centimeters. Under these conditions, the output signal from the photomultiplier shows a complex pattern, caused by the different propagation delays of the light coming from each illuminated leaf (See Fig 4). The situation is further complicated if we consider that the laser spot can be intercepted by non-fluorescent materials such as stems or flowers, which give an important contribution to the backscattered signal. We also have to consider that the backreflectance contains an important specular component, which is not present in the fluorescence emission. An additional decorrelation would then be introduced by the different inclinations of leaves inside the canopy.

For studying the interaction between a short laser pulse and the canopy, we have performed measurements on an artificial tree made of paper leaves impregnated with oxazyne, a fluorescent dye (see Fig 4).

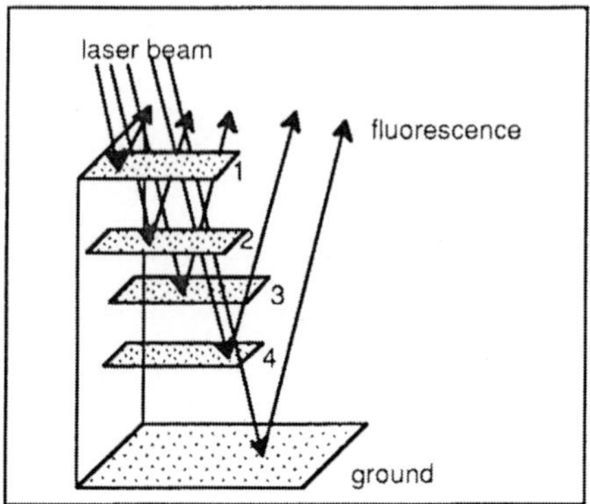

<u>Fig 4:</u> Propagation of ligth in an artificial tree. The fluorescence emission coming from deeper layers are delayed compared to those coming from the front layer.

Fig 5 shows the time variation of the output signal when this tree is illuminated with the laser spot. Several peaks appear, each corresponding to the fluorescence emission from a different leaf. The backscattered signal, at the same wavelength as the excitation beam, shows also different peaks of light coming from different leaves. It also shows a contribution of the ground level, which is illuminated by the laser spot (see Fig 4). To extract the fluorescence lifetime information from these complex signals, we modelized the backscattered function of the canopy, making the assumption that the contribution of each leaf is equivalent to those of a single plane surface perpendicular to the laser beam. This backscattered function of a single surface is recorded separately. The relative positions and the amplitudes of the peaks are determined by a least squares method using the Marquardt algorithm. Table I shows the results of this calculation compared to the actual distances measured on the artificial tree. These values indicate that our model is able to represent the data. It also shows that the leaf distances distribution of a canopy, which is an important structural parameter, can be obtained from the backreflected signal alone.

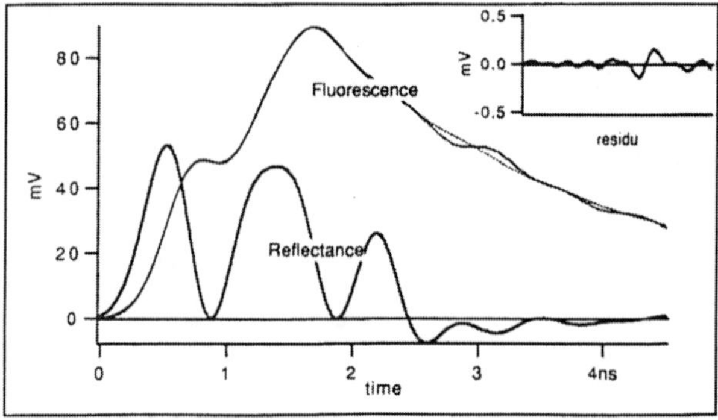

<u>Fig 5:</u> Fluorescence and reflectance decay from the arficial tree shown in Fig 4.

On the basis of this decomposition of the backreflected signal, the fluorescence signal can now be calculated, making the assumption that each leaf has the same fluorescence lifetimes (i.e., the same fluorescence decay F(t)), but the emission amplitude of each leaf is allowed to vary from one leaf to another. The search for the fluorescence parameters (lifetimes and fluorescence amplitude coefficients for each leaf) is made using the already mentioned Marquardt algorithm. Fig 5 shows that with a monoexponential decay having a lifetime of 2.4 ns, the recalculated signal Fc(t) is close to the experimental decay Fe(t). This lifetime value is similar to the 2.3 ns lifetime that has been measured on a dry deposit of oxazyne on a filter paper.

	leaf 0	leaf 1	leaf 2	leaf 4	ground
actual position (cm)	0	5	10	15	25
calculated position (cm)	0	-	9.5	14.8	24.9
amplitude of backreflectance	30	0	16	28	15
amplitude of fluorescence	31	0	40	25	2

Table I: Retrieved relative positions of the leaves of the artificial tree presented in Fig 4. The contribution of leaf 1 is zero, so its position cannot be calculated. Backscattering and fluorescence characteristics of the leaves are also given.

We can conclude that, thanks to our calculation method, fluorescence lifetime information can be extracted from the signals produced by a complex target. This method is also applicable in the case of a real canopy like sorghum, as shown in Fig 6. The backscattered signal has been interpreted as a sum of six elementary pulses produced by six different layers of the canopy, whose positions are indicated in Fig 6. The fluorescence signal has been interpreted as a biexponential decay, with an average lifetime of 0.32 ns. Other canopy type such as wheat or soja has been tested successfully (data not shown).

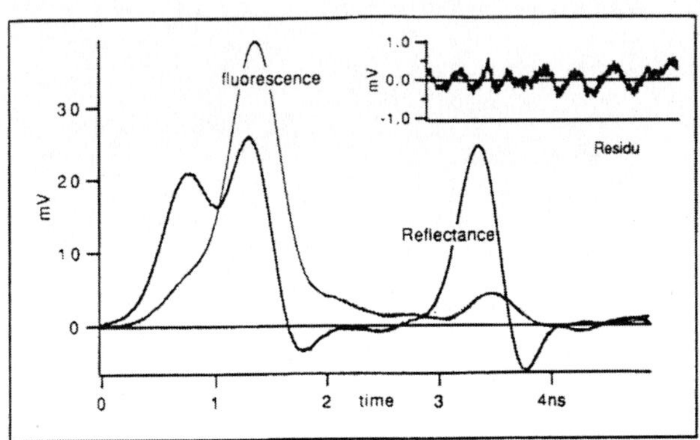

Canopy layer n°	1	2	3	4	5	6
Calulated position above ground (cm)	44.7	40.3	37.1	31.8	28.1	0.0
Amplitude of reflectance	1	10.3	9.3	18	9.5	19
Amplitude of fluorescence	5	4.8	4.7	19.5	15.6	1.9

Fig 6: Fluorescence and reflectance signal from a sorghum canopy. The signal is 16 times accumulated. The figure shows the complex pattern due to the particular fluorescence and

reflectance properties of each layer. The recalculated fluorescence is closed to the experimental decay, as indicated by the residue function.

Conclusion

The results presented here show that, with a special method of deconvolution, remote sensing of the fluorescence lifetime can be performed on plant canopies with laser instrumentation. They also show that information about the relative position of leaves inside the canopy can be extracted from the backscattered signal only. This information would provide data to canopy models developed for the understanding of interactions between vegetation and its environment. Fluorescence lifetime measurement on canopy offers new perspectives for vegetation monitoring at a large scale, because it is closely related to photosynthetic activity. Experiments are now being performed with this new type of picosecond LIDAR and are mainly focused on developing methods for assessing plant status and stress.

[1] Krause, G.H., E. Weis: "Chlorophyll fluorescence and photosynthesis: The basics", Annu Rev Plant Physiol Plant Mol Biol, 42, 313-349 (1991).

[2] Björkman, O., B. Demmig: "Photon yield of O_2 evolution and chlorophyll fluorescence characteristics at 77K among vascular plants of diverse origins", Planta, 170, 489-504 (1987).

[3] Weis, E., J. Berry : "Quantum efficiency of photosystem II in relation to 'energy'-dependent quenching of chlorophyll fluorescence", Biochim Biophys Acta, 894, 198-208 (1987).

[4] Duysens, L., H. Sweers: "Mechanism of the two photochemical reactions in algae as studied by means of fluorescence", in: Jap. Soc. of Physiol. Studies on microalgae and phosynthetic bacteria, Univ. of Tokyo Press, Tokyo, (1963).

[5] Genty, B., J. Briantais, N. Baker: "The relationship between the quantum yield of photosynthetic electron transport and quenching of chlorophyll fluorescence", Biochim Biophys Acta, 990, 87-92 (1989).

[6] Zimmermann, R., K. Günther: "Laser-induced chlorophyll-a fluorescence of terrestrial plants.", Proc IGARSS'86 (Zürich), Vol III, 1609-1613 (1986).

[7] Moya, I., Y. Goulas, J. Briantais: "Techniques pour la télédétection de la durée de vie et du rendement quantique de la fluorescence de la chlorophylle in vivo", Proc 4th Int Coll on Spectral Signatures of Objects in Remote Sensing (Aussois (France)), (1988).

[8] Moya, I., M. Hodges, J. Barbet: "Modification of room-temperature picosecond chlorophyll fluorescence kinetics in green algae by photosystem II trap closure", FEBS lett, 198, 256-262 (1986).

[9] Goulas, Y.: "Télédétection de la fluorescence des couverts végétaux: Temps de vie de la fluorescence chlorophyllienne et fluorescence bleue", Thesis, Université de Paris-Sud, (France), (1992).

[10] Marquardt, D.: "An algorithm for least-squares estimation of nonlinear parameters", J. SIAM, 11, 431-441 (1963).

[11] Nash, J.: "Minimising a nonlinear sum of squares", in: Compact numerical methods for computers: linear algebra and function minimisation, A. Hilger, Bristol, New York, 207-217 (1990).

[12] Morales, F., A. Abadia, J. Abadia: "Chlorophyll fluorescence and photon yield of oxygen evolution in iron-deficient sugar beet (*Beta Vulgaris* L.) leaves", Plant Physiol, 97, 886-893 (1991).

Laser Induced Chlorophyll Fluorescence Induction Kinetics as a Tool for the Determination of Herbicide Action in Algae

B. Ruth

GSF - Research Centre for Environment and Health, Institute of Soil Ecology, 85758 Neuherberg, Germany

Herbicides are widely applied in agriculture. In order to assess the ecological consequences of their use, the action on those plants must be also considered which are not the target of the application. These plants are subjected to relatively low herbicide concentrations. Herbicides are washed from the fields into creeks and ponds, where they act on algae and other plants. Measurements on algae have the further advantage to allow the application of herbicides of a given concentration without difficulties arising from restricted uptake in plants.

Herbicides of the DCMU-type such as atrazine or terbuthylazine block the action of the secondary electron acceptor in the electron transport chain [1]. In this way, the effectivity of the photosynthetic system to process the absorbed light is significantly reduced. Measurements of the chlorophyll fluorescence induction kinetics provide one way to evaluate the effectivity of different components of the photosynthetic system [2].

The chlorophyll fluorescence induction kinetics can be observed after the plants have been adapted to darkness for at least 15 - 20 min. The excitation with light of constant intensity induces the fluorescence intesity to rise within microseconds to an initial value F_O. Then the fluorescence continues to rise with different time constants to the peak value F_p and declines again to the steady-state value F_s within about 5 min. The rise from F_O to F_p may also include one or more inflection points or an intermediate maximum and minimum [3]. Although this feature of the induction kinetics can be observed from all plants, the details of the tracing and especially the intensity depend on e.g. the considered plant, chlorophyll content, pre-illumination, period of dark adaptation, and temperature [2].

F_O represents the fluorescence intensity of the antenna without any influence of the red-ox state of the following electron transport chain. The increasing reduction of the electron acceptors causes the fluorescence intensity to increase from F_O to F_p because the electrons cannot be processed and the energy must be emitted partly by the fluorescence light.

As the DCMU-type herbicides block the electron acceptor Q_b, the fluorescence intensity shows an additional increase when compared to the control [4]. However, the fluorescence intensity depends also on other conditions e.g. the chlorophyll content.

It is therefore the aim of this paper, to derive a parameter
from the induction kinetics which scales directly the herbicide
action. In order to separate the primary effects of the
herbicide action from other consequences due to indirect
effects, the time resolution must be sufficient to follow the
time constants related to the electron acceptors Q_a and Q_b.

The algae are adapted to darkness in a small vessel (0.5 cm^3)
in a measuring head for 15 min. An optical fibre bundle guides
the excitation light of a He-Ne-laser to the algae using an
accousto-optical modulator as a high-speed shutter. In this
way, the excitation light (8 mW/cm²) is switched on within less
than 1 μs. A further optical fibre bundle collects the
fluorescence light selected by a filter combination at 685 nm.
An electronic device samples the fluorescence intensity with a
maximum time resolution of 10 μs. For the slow part of the
induction kinetics the time resolution is reduced. A computer
stores the data and controlls the acousto-optical modulator
[5].

Different representatives for green algae (Scenedesmus
quadricauda), blue algae (Microcystis aeruginosa), and diatoms
(Navicula pelliculosa) were used for the measurements. They
were cultivated in ventilated nutrient solution in 2 l flasks
with permanent stirring. The pre-illumination of white light
had an intensity of 2 mW/cm².
One day before the measurement, they were diluted to a density
characterized by an absorption coefficient between 0.2 and 0.3
at 750 nm.

As representative for the herbicides, terbuthylazine (TA) was
chosen. Commercially available Gardoprim 500^R with a TA-
concentration of 490 g/l was diluted to get a stock solution of
1 mg/l. After the algae had been measured without herbicides as
a control, this solution was added to the algae to obtain a
final concentration in the range 5 μg/l - 200 μg/l.

As the induction kinetics is measured with a maximum time
resolution of 10 μs and with a total measuring time of 5 min,
the details of the curve can only be explained in one curve if
the time axis is logarithmic. Fig. 1 (curve 0) shows as control
the induction kinetics of scenedesmus without any effect of the
herbicide. One can clearly identify the initial intensity F_o,
the peak intensity F_p, and the final steady-state value F_s.

As preferably applied in the forest decline research, the ratio

$$R_{fd} = (F_p - F_s) / F_s \qquad (1)$$

is regarded as a measure of the potential photosynthetic
activity [6]. The curves indicated by 10, 20, 50, 100, and 200
represent the induction kinetics obtained with the
corresponding TA-concentrations given in μg/l. They show that
R_{fd} decreases clearly for the concentrations 100 and 200 μg/l.
In contrast to the control, these curves increase again after
the intermediate decline following F_p in order to obtain the
final steady-state value F_s. This makes the interpretation
difficult.

As shown by the curves in Fig. 1, an additional shoulder in the range 100 μs - 100 ms is developping with increasing TA-concentration. This time range is clearly related to the action of the secondary electron acceptor Q_a. The consideration of this feature in the curve has the advantage that it does not depend on the absolute fluorescence intensity. In order to quantify this shoulder, the fluorescence intensities $F_1(t_1)$, $F_2(t_2)$, and $F_3(t_3)$ are derived and the ratio

$$B = (F_2(t_2) - F_1(t_1)) / (F_3(t_3) - F_1(t_1)) \qquad (2)$$

is calculated with the times t_1 = 316 μs (log t_1/s = -3.5), t_2 = 3.16 ms (log t_2/s = -2.5), and t_3 = 31.6 ms (log t_3/s = -1.5). B scales the relative increase of F(t) in the first 10 % of the given time interval.

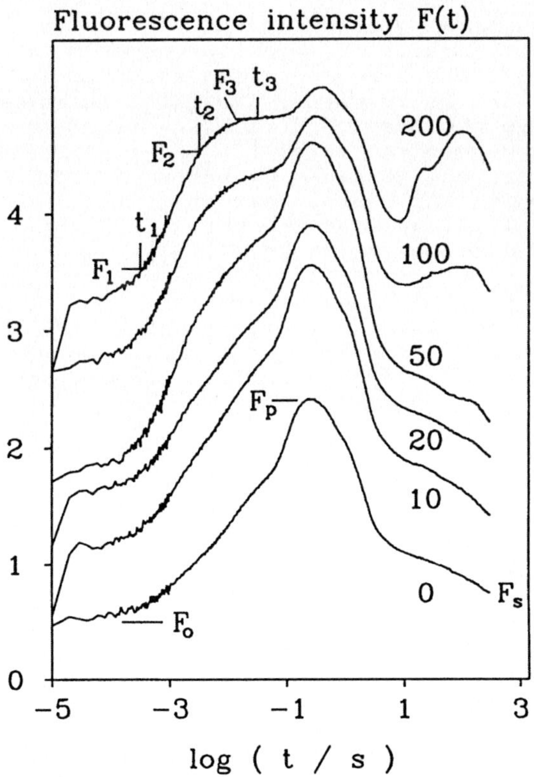

Fig. 1 Fluorescence induction kinetics obtained from the algae Scenedesnus with an terbuthylazine concentration of 50 μg/l. The curves are indicated by the concentrations given in μg/l.

The time dependence of B after the application of terbuthylazine of the concentration 50 μg/l to the three species of algae shows a first increase after 20 min and an approximation

to the steady state 60 - 100 % above the control after 3-6 h.
The control values are constant within 10 %. In comparison to
the control, the R_{fd}-values decrease between 11 % and 50 % and
they achieve their steady-state effect within 3 h. However, the
control values of R_{fd} itself have a relative change of more
than 50 % during the measurement of 6 h. This comes presumably
from the hour of the day.

Parameter B shows an significant increase after the application
of TA-concentrations of 5 μg/l for all three types of algae.
This demonstrates the sensitivity of parameter B to determine
the herbicide action. B is further increased for rising
concentrations until it approaches the saturation value about
100 % above the control value for a concentration near 200
μg/l. The R_{fd}-values show the first effects above 20 μg/l and
they decline between 50 % and 80 % relative to the control
value.

The concentration of 5 μg/l, at which the first effects can be
observed with the parameter B, must be compared to typical
concentrations in the environment. The recommended
concentration of the commercially available herbicides is about
1 kg/ 10000 m². In the case of a rainfall of 10 mm, this
corresponds to a concentration of 10 mg/l if the absorption
effects in soil are not considered. Water of this concentration
can flow into creeks and ponds. Rainwater itself can have a
concentration of 1.23 μg/l and the total wet deposition of
terbuthylazine can amount to 11 μg/m² [7].

The results show that the initial phase of the induction
kinetics can provide useful information about the action of
herbicides in concentrations relevant for the environment.

REFERENCES

[1]   Böger P & Sandmann G (1989) Target sites of herbicide
action, CRC press, Boca Raton, Florida, USA.
[2]   Krause G H & Weis E (1984) Chlorophyll fluorescence as a
tool in plant physiology. II. Interpretation of fluorescence
signals, Photosynthesis Research **5**, 139-157.
[3]   Govindjee, Amesz J & Fork D C (Edts) (1986) Light emission
by plants and bacteria, Academic Press, Orlando, Florida, USA.
[4]   Voss M, Renger G, Kötter C & Gräber P (1984) Fluorometric
detection of photosystem II herbicide penetration and
detoxification in whole leaves, Weed Sciences **32**, 675-680.
[5]   Ruth B (1990) A device for the determination of the
microsecond component of the in vivo chlorophyll fluorescence
induction kinetics, Measurement, Sience & Technology **1**, 517-
521.
[6]   Lichtenthaler H K (Edt.) (1988) Applications of
chlorophyll fluorescence, Kluwer Academic Publishers,
Dordrecht, The Netherlands.
[7]   Oberwalder Ch, Gießl H, Irion L, Kirchhoff J & Hurle K
(1991) Pesticides in rainwater, Nachrichtenblatt Deutscher
Pflanzenschutzdienst **43**, 185-191.

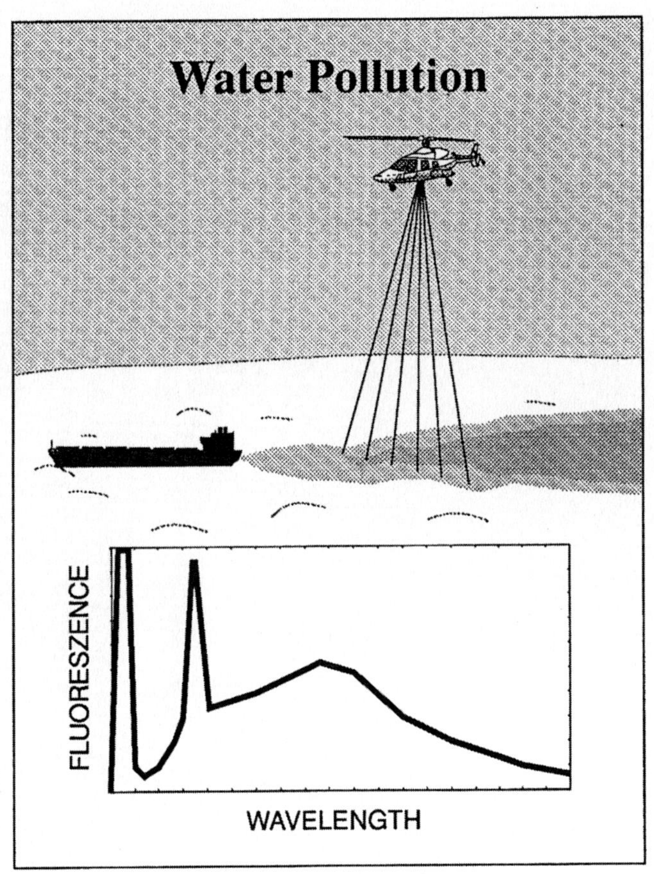

Water Pollution
FLUORESZENCE
WAVELENGTH

Lidar Monitoring of Tyrrenian and Northern Adriatic Seas

Giovanna Cecchi, Marco Bazzani, Luca Pantani
I.R.O.E.- C.N.R., Via Panciatichi 64, I 50127 Firenze, Italy

Valentina Raimondi
Università di Firenze, Dept.Electronic Engineering, Via S.Marta 3, I 50100 Firenze, Italy

The protection of the marine environment from pollution is a matter of major importance. Remote sensing techniques provide non-intrusive measurements and coverage of large areas and optical sensors are particularly suitable because the visible radiation is the only part of the electromagnetic spectrum which shows a good penetration in the water column. Passive optical sensors have problems in coastal waters where a large quantity of suspended sediments and dissolved substances are present (class 2 waters), while fluorescence lidars [1] (active remote sensing) give a more detailed assessment of the water properties and allows the detection of parameters which can not be obtained by passive analysis. Lidar techniques have been used for oil spill detection and algae studies for quite a long time [2], but in the second half of the eighties the improvement of detector arrays allowed the development of high spectral resolution fluorosensors and therefore the introduction of new processing techniques for the analysis of marine parameters like water column temperature [3] and phytoplankton[4].

The FLIDAR family(*)[5], developed at IROE-CNR in Firenze from 1985, was the first high spectral resolution fluorosensor with a low weight and compact structure which can operate as an airborne or surface-based sensor. In the present paper field experiments carried out in marine environment with the last prototype, FLIDAR-3, are reported. The main purpose of these experiments was to test the potential of high spectral resolution fluorosensors in remote measurements of the water parameters. A standard set of measurements was established as follows:
1 - 308 nm excitation, 2400 g/mm grating, centre wavelength 344 nm (water Raman)
2 - 308 nm excitation, 150 g/mm grating, centre wavelength 450 nm (oil and DOM)
3 - 308 nm excitation, 150 g/mm grating, centre wavelength 650 nm (phytoplancton)
4 - 480 nm excitation, 150 g/mm grating, centre wavelength 650 nm (phytoplancton)
but in some of the experiments only part of the set was used.

Four main experiments were carried out in different locations an from different platforms:

THE MARET-91 EXPERIMENT - This campaign was held in the Upper Tyrrenian Sea in September 1991. The lidar van was placed on the rear deck of the O.S.Minerva and the laser beam was brought perpendicular to the water surface by means of a 45° mirror. Complete sets of measurements were carried out 24 hours a day at regular intervals. Extra sets were carried out when the ship stopped for *in situ* measurements. The main purpose of the campaign was the test of a high spectral resolution lidar in the detection of marine parameters with a

---

(*)   FLIDAR is an international trademark of CNR

particular emphasis on phytoplancton detection and identification [4] and in the measurement of water column temperature by Raman techniques [3].

As a general consideration very low presence of oil and chlorophyll was detected over all the path, this fact was confirmed by *in situ* measurements. Therefore it was impossible to fully test the FLIDAR-3 performances in the detection and identification of phytoplancton. Hopefully the campaign was held in correspondence of a bloom of "mucilagine" in the Tyrrenian waters. Mucilagine is a not well understood phenomenon consisting in the presence inside the water column of organic jelly-like substances, probably produced by phytoplancton under particular meteorological conditions. At the beginning of September the mucilagine was distributed in the water column and was not detectable by passive observation of the water surface. The 308 nm excitation gave a quite structurated and characteristic spectrum with a wide band around 475 nm and a low chlorophyll band at 685 nm, due to the small quantities of phytoplankton encapsulated inside the mucilagine. Similar spectra were detected along all the ship travel indicating that the phenomenon covered at least all the central part of the Thyrrenian Sea. While the surface, temperature is easily detectable by infrared passive sensors the detection of water Raman spectra with high spectral resolution is the only technique which allows the remote sensing of water column temperature. One of the purposes of the campaign was the test of a new procedure [3] for the extraction of water column temperature from the Raman signal. The Raman spectra were deconvolved in term of Gaussian components and the ratio between the amplitudes of the two main Gaussian curves was compared with the water column temperature measured at the ship stops. The regression curve was very linear with a reproducibility better than 0.06 °C.

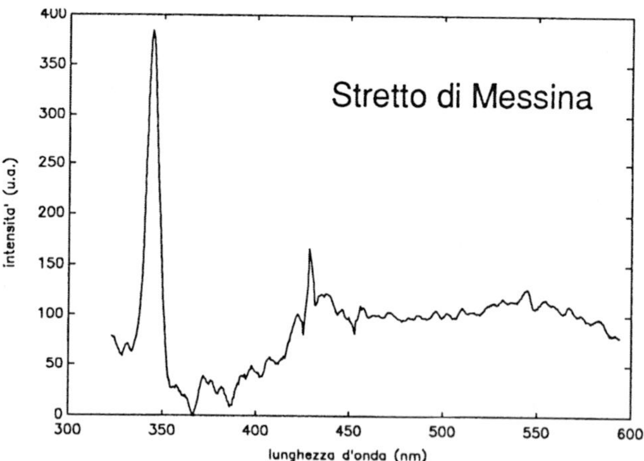

Fig.1 - LIRA-92 EXPERIMENT: Raman and Fluorescence spectrum.

THE LIRA-92 EXPERIMENT - This campaign was held in the Sicily waters on March 1992 in cooperation with the Engineering faculty of Catania University [6]. The FLIDAR-3 was installed inside an AB-212 antisubmarine helicopter of the Italian Navy over the trap of the sonar buy. The main purpose of the campaign was the test of the FLIDAR-3 in airborne operations with a particular attention to the detection of oil spills and their discrimination from other organic substances; as a consequence only item 2 of the measurement set was used.

The field investigation was conducted over a study area including part of the eastern Sicilian coastal zone, having a high risk of oil pollution. This included the Augusta bay, the Messina strait, the Milazzo gulf, the Gela gulf and off-shore oil platforms. The measurements were done at different altitudes and speeds in order to have a full test of the fluorosensor operativness. Weather during the flights was characterized by a considerable variability that made it possible to verify the FLIDAR-3 operativness under different conditions. The carrier vibrations did not disturb the fluorosensor operations and no electromagnetic interference between the fluorosensor and the on board electronics was observed during the flights. Different situations were detected ranging from heavy oil pollution in Augusta bay to quite clear water in the Milazzo gulf. In Augusta and the Messina Channel the oil was present together with the DOM and the experiment clearly demonstrated the usefulness of a high spectral resolution in distinguishing between the two.

THE SILAV-92 EXPERIMENT - The Venice Lagoon is a particular environment which is critical for passive optical sensors because of its class 2 waters and for the interference of the lagoon bottom due to the very low bathimetry. In this situation the use of fluorescence lidar is very attractive. In the frame of the CNR Project *"Salvaguardia del sistema lagunare Veneto"* experiments were successfully carried out in order to test the usefulness of fluorescence lidar remote sensing in the monitoring of lagoon processes.

In July 1992 a first test was done from the pier in Punta Sabbioni in front of one of the *"bocche"* which connect the lagoon with the open sea. Two fluorescence lidars, one from Lund Institute of Technology and the FLIDAR-3, were operated while *in situ* measurements of water quality parameters were carried out by a research vessel of Istituto per lo Studio della Dinamica delle Grandi Masse. Raman and fluorescence data on turbidity, dissolved organic matter, and chlorophyll were recorded for 24 hours and compared with the corresponding *in situ* data. In August 1992 a two days campaign was carried out using the CNR O.S.D'Ancona, devoted to *in situ* measurements, and a small boat, carrying the FLIDAR-3 van. Raman signal, chlorophyll and DOM fluorescence were detected along a path which included all the most significant parts of the lagoon waters and an agreement was found between lidar and *in situ* measurements.

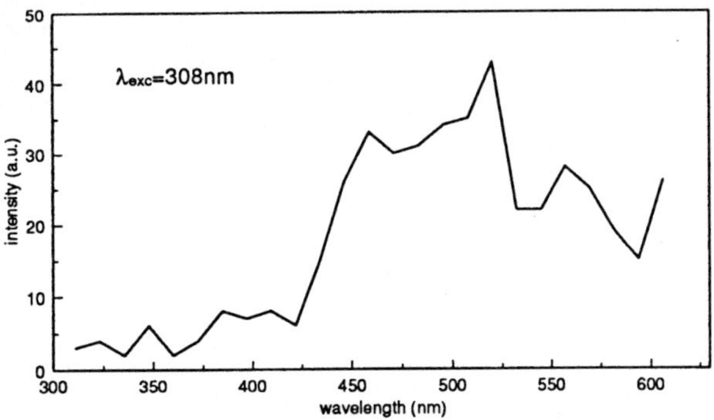

Fig.2 - LARA-93 EXPERIMENT: Raman and fluorescence spectra.

THE LARA-93 EXPERIMENT - The last experiment was carried out over the Po delta and the Venice lagoon in April 1993. The FLIDAR-3 was installed inside the CASA 212 aircraft of the CNR LARA project. The main purpose of the experiment was to test the operativness of the FLIDAR-3 as an airborne sensor. The full set of measurements was successfully tested from different altitudes, yellow substances and chlorophyll were detected. The use of a suitable channel grouping allowed the detection of spectra also from a 1000 m altitude.

Also if a deeper analysis of the data is necessary some relevant results were obtained in these field experiments:
1 - The high resolution lidar was shown to be operative from different platforms included aircrafts and helicopters also from quite relevant altitudes,
2 - The high resolution lidar was able to detect mucillagine also if distributed in the water column and therefore not detectable by other remote sensors,
3 - A new technique for the remote sensing of water column temperature was successfully tested which allows an high accuracy and sensitivity.

REFERENCES

1] MEASURES R.M. (ed.) (1988):"Laser Remote Chemical Analysis" (New York: Wiley-Interscience).
2] PANTANI L., CECCHI G., (1990):"Fluorescence lidars in environmental remote sensing" in MARTELLUCCI, S., and CHESTER, A.N., (eds.) Optoelectronics for Environmental Science, (New York: Plenum Press), 131-148.
3] BAZZANI M., BRESCHI B., CECCHI G., PANTANI L., TIRELLI D., VALMORI G., CARLOZZI P., PELOSI E., TORZILLO G. (1992) "Phytoplankton monitoring by laser induced fluorescence", EARSeL Advances in Remote Sensing, 1, 106-110,
4] BRESCHI B., CECCHI G., PANTANI L., RAIMONDI V., TIRELLI D., VALMORI G., MAZZINGHI P., ZOPPI M.:"Measurement of water column temperature by Raman scattering", EARSeL Advances in Remote Sensing, 1, 131-134, 1992
5] CECCHI G., PANTANI L., BRESCHI B., TIRELLI D., VALMORI G. (1992) "FLIDAR: a multipurpose fluorosensor-spectrometer", EARSeL Advances in Remote Sensing, 1, 72-78,
6] GERACI A.L., LANDOLINA F., CECCHI G., PANTANI L. (1993):"Laser and infrared techniques for water pollution control", Proc.1993 Oil Spill Conference, 525-529, St.Louis

Investigations on the Photosynthetic Activity of Cyanobacteria of the Baltic Sea Using a Mobile Picosecond-Fluorimeter

Karlheinz Maier-Schwartz, Frank Terjung, Dirk Otteken (FB Physik),
Ulrich Fischer, Beate Meyer, Jörg Rethmeier (FB Biologie and ICBM)

Carl-von-Ossietzky Universität, POB 2503, 26111 Oldenburg, Germany

Pollution and eutrophication of marine waters continue being important environmental problems. High amounts of sulfide and therefore a deficiency of oxygen are found in shallow coastal regions of the Baltic Sea. Some cyanobacteria species can survive under such sulfidic conditions and still introduce oxygen into the sediment and the overlying water thus improving environmental conditions. These organisms are either tolerant of high sulfide concentrations or adapt their photosynthesis by switching from an oxygenic type to an anoxygenic one. Different cyanobacteria were isolated from the island of Hiddensee (southern Baltic Sea) and used for laboratory experiments. Using our recently developed mobile picosecond-fluorimeter [1] we observed variations in the photosynthetic activity of pure cultures of these cyanobacteria due to their response to different sulfide concentrations in the medium. The analysis of the data revealed a correlation of the fluorescence decay to the adaption of the cyanobacteria to sulfide stress and to the production of oxygen. In a first *in situ* experiment the device was proved at the Zwischenahner Meer (Fig. 1).

Figure 1: *In situ* application of the mobile picosecond-fluorimeter for the measurements of photosynthetic activity of cyanobacteria and green algae at the Zwischenahner Meer.

106

The primary reactions of photosynthesis in cyanobacteria can be sketched in the so-called Z-interaction-scheme. Light is absorbed by the pigment molecules phycoerythrin, phycocyanin, allophycocyanin and chlorophyll and transferred to the reaction centers of photosystem II and photosystem I. Under aerobic conditions water serves as electron donor to reduce reaction center II. Simultaneously free oxygen is liberated. From the optically excited center II an electron moves via a long chain of electron transfer molecules to reduce center I. When the photosystems are excited by an ultrashort laser pulse the absorbed energy is channelled to photochemistry and partially to the production of heat and to fluorescence. In a simple relaxation rate model this results in a fluorescence decay given by a sum of exponentials

$$F(t) = \sum_i \alpha_i \exp\left(-t/\tau_i\right)$$

In numerical analysis of decay curves a wide variation of these individual decay components is found and the fit parameters are rather difficult to interprete.

For practical applications it is more convenient to calculate the average decay time

$$\tau_m = \frac{\int\limits_0^\infty t\, F(t)dt}{\int\limits_0^\infty F(t)dt} = \frac{\sum\limits_i \alpha_i \tau_i^2}{\sum\limits_i \alpha_i \tau_i}$$

As shown previously, this can be an appropriate parameter to characterize the activity of photosynthesis for a given organism with two photosystems [1] . The fluorescence decays of pure cultures of cyanobacteria and of green algae are shown in Fig. 2 . The average decay times are 338 ps and 754 ps, respectively. These average decay times proved to be representative of those species. In the *in situ* experiments (see Fig. 1) we measured nearly identical decay curves as shown in Fig. 2 for different parts of a microbial mat. A subsequent microscopic analysis confirmed our classification of cyanobacteria and green algae.

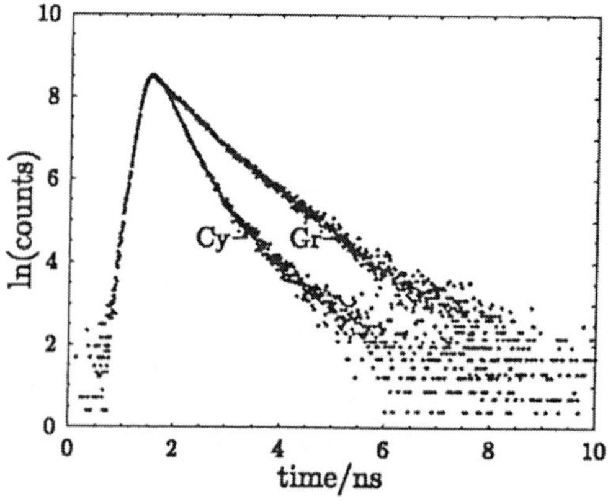

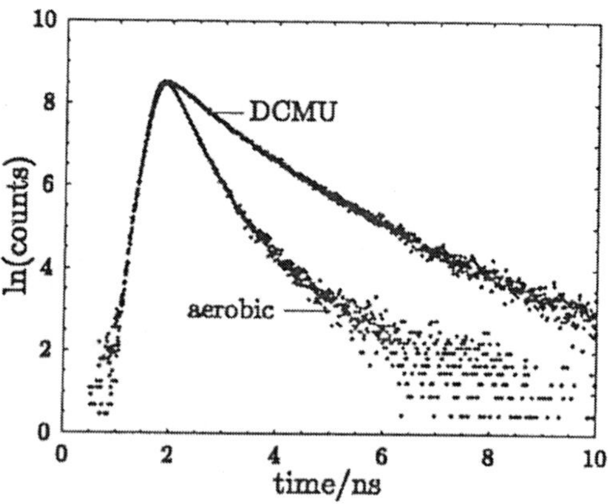

Figure 2: Logarithmic plot of the fluorescence decay of the cyanobacterium *Oscillatoria spec.* (Cy) and the green alga *Eremosphaera viridis* (Gr) under aerobic conditions

Figure 3: Logarithmic plot of the fluorescence decay of the cyanobacterium *Oscillatoria spec.* under aerobic conditions and in the presence of DCMU

The herbicide DCMU blocks the electron transport chain between the photosystems II and I very efficiently, which prevents the reaction centers of type II from being active in electron transfer any longer. They are closed and the production of oxygen is stopped. This decrease of photochemistry reduces the sum of the relaxation rates and leads to a slower decay of the excited photosystems II and correspondingly their fluorescence. Fig. 3 shows the effect of a total blocking of photosystems II of cyanobacteria due to a sufficiently high concentration of a DCMU solution. The average decay time enlarges from 338 ps to 1024 ps. As long as there is no destruction of physiological structures these two curves represent extreme situations. Any variation of photosynthetic activity caused by stress effects should give results between.

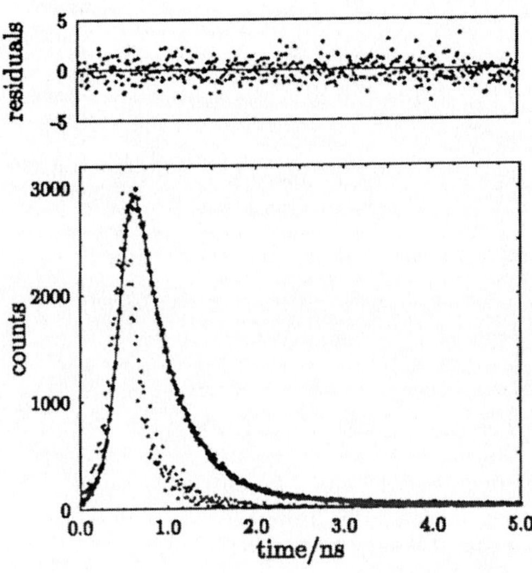

Figure 4: Linear plot of the fluorescence decay of the cyanobacterium *Synechocystis spec.* and of a cyanin fluorescent scatterer. The full curve represents the fit. The randomly distributed weighted residuals confirm the quality of the fit.

Our experimental set-up consists of a red laser diode (652 nm, 300 ps FWHM, 500 kHz repetition rate), two optical fibers (4 m length) with a special detector head for excitation pulse and fluorescence transmission, an interference filter at photosystem II emission wavelength (695 nm), a time correlated photon counting system and a 486 - based computer with an integrated pulse-height-analysis DMA-card. The apparatus function is obtained by measuring the fluorescence decay of a well-known cyanin dye fluorescent probe (about 50 ps). Data analysis is carried out with standard deconvolution techniques following a chisquare criterium for quality test of the fit. Fig. 4 shows a typical three exponential fit of a decay curve of the cyanobacterium *Synechocystis spec.* under sulfide stress. The corresponding parameters are $\{\alpha_1, \alpha_2, \alpha_3\} = \{0.75, 0.22, 0.03\}$ and $\{\tau_1, \tau_2, \tau_3\} = \{158\ ps, 380\ ps, 1631\ ps\}$ with an average decay time of 519 ps. The weighted residuals and the chisquare of 1.04 are of good quality.

We have made numerous investigations with pure cultures of cyanobacteria isolated from the Baltic Sea exposed to various sulfide concentrations of up to 5 mM. Since different reactions of the organisms to the sulfide stress were found they can be divided into four types:
- species which do not survive, the fluorescence decay curves partially contain long-lived components due to released pigments,
- species which are slightly tolerant of sulfide stress, the decay time remains rather unchanged,
- species which create themselves a protection mechanism against sulfide toxification, and
- species which reversibly adapt their photosynthesis to sulfide stress.

An example for a protection mechanism is shown in Fig 5. Under sulfidic conditions the cyanobacterium *Oscillatoria spec.* forms spherical aggregates in which the photosynthetic activity of the organisms at the outside is strongly inhibited, whereas the organisms in the inside are protected and show a behaviour close to unstressed ones.

This means that inside this aggregate aerobic conditions are predominant. Oxygen production is still performed. The corresponding decay times are 853 ps and 407 ps, respectively.

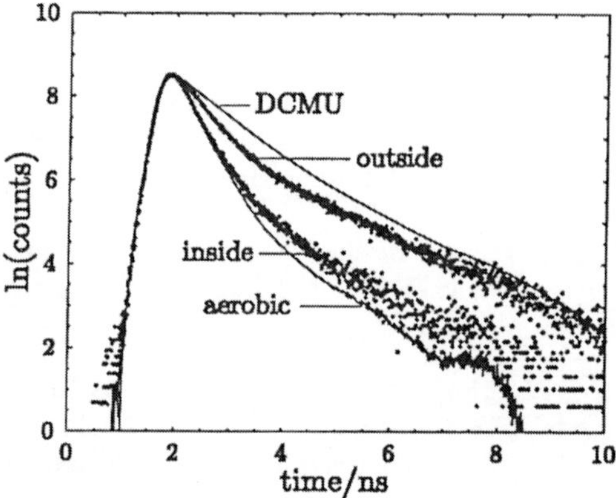

Figure 5: Logarithmic plot of the fluorescence decay of the cyanobacterium *Oscillatoria spec.* under sulfidic conditions. The dotted curves are monitored outside and inside the aggregate. The full curves represent the fits of the extreme conditions.

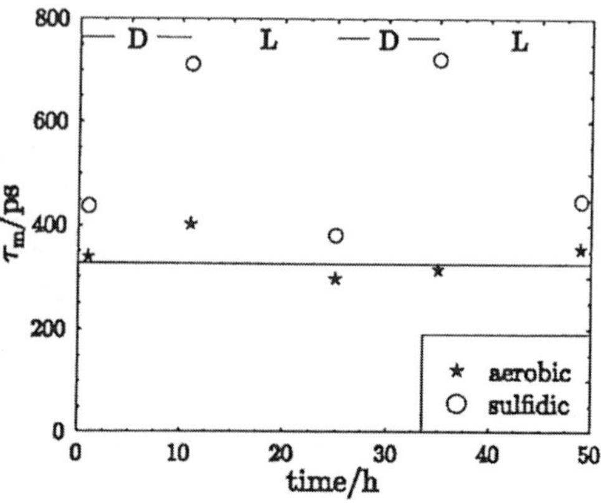

Figure 6: Average decay times of the cyanobacterium *Anabaena variabilis* as calculated from data taken after light (**L**) and dark (**D**) periods. Cultures under sulfidic and for control aerobic conditions.

The influence of a dark / light cycle on the fluorescence decay of the cyanobacterium *Anabaena variabilis* in the presence ór absence of sulfide (1.2 mM) is shown in Fig 6. The cultures are illuminated with moderate light intensities during light periods. As can be seen from Fig 6, there is no significant difference in the photosynthetic efficiency in unstressed organisms independently of illumination conditions. The sulfide stressed culture has a photosynthetic activity close to the unstressed organisms but only after longer periods of illumination. Photosynthetic activity in the presence of sulfide was low after dark periods and increased upon illumination. Apparently, this cyanobacterium uses sulfide as additional effective electron donor for its photosynthesis.

From these experiments it becomes evident that our mobile picosecond-fluorimeter is well suited for detailed studies of marine photosynthetic organisms and their environmental conditions.

Acknowledgement: This study was partially supported by a grant from the Bundesministerium für Forschung und Technologie (DYSMON-Projekt) to U.F.

[1] K. Maier-Schwartz, E. Breuer, H.G. Hegeler
Mobile Picosecond-Fluorimeter for Studying Vegetation Stress
in "Laser in der Umweltmeßtechnik", ed. C.Werner, V. Klein, K.Weber,
Springer-Verlag , Berlin 1992, pages 70-73

UV Laser Induced Fluorescence to Determine Organic Pollutions in Water

G. Hillrichs, W. Neu
Laser-Laboratorium Göttingen e.V.
Im Hassel 21, 37077 Göttingen

Abstract

UV-laser induced fluorescence is a sensitive method to detect pollution of water by a variety of organic compounds, especially by various oil products. The combination of lasers with fiber optics allows an in situ detection without sampling. However, for short wavelengths in the UV spectral range the laser radiation suffers high basic attenuation in the fiber material. Additionally, laser induced effects increase the absorption of the fused silica fiber. In this work, we studied these effects and the resulting limitations for the fiber optical sensor. We give some examples for detection of several water pollutants at various laser wavelengths and for different water qualities.

Introduction

The detection of water pollution by different organic compounds using laser fluorimeters in combination with fiber optical sensors has become a well studied technique during the last years [1-4]. Especially, detection of contamination by oil and different oil products is possible. If remote detection of the water quality is necessary for longer times the fluorimetric methods based on fiber optics can be a cost effective alternative for other well established methods based on probe sampling and laboratory probe analysis. However, there are some constraints for the selectivity and sensitivity of the laser fluorimetric method.

The construction of a mobile laser fluorimeter for remote sensing requires the availability of small compact laser sources in the UV and of optical fibers which allow transmission of the laser light and of the induced fluorescence over long distances. The nitrogen laser ($\lambda = 337$ nm) has become a well studied source for a UV laser fluorimeter [2-4]. However, the range of detectable pollution species and also the selectivity and the sensitivity of the fluorimeter could be improved, if other additional shorter excitation wavelengths would be available. Especially small aromatic compounds which are contents of gasoline (i.e. benzol and toluene) absorb preferentially at shorter wavelengths. Possible excitation sources are frequency upconverted solid state lasers (266 nm [1], 355 nm) and compact versions of excimer lasers (248 nm, 308 nm). For several of these wavelengths we studied the suitability for pollution detection.

Contrary to the visible or the near infrared spectral region transmission of UV light is strongly limited by increased fiber attenuation and also by laser induced effects in the fiber core. We will discuss some details about the fiber behaviour in the UV range.

UV Lasers and Fiber Optics

The most suitable excitation source will be a continuously tunable laser covering the spectral region from about 240 nm to 400 nm. Up to now only frequency doubled dye lasers can fullfill these requirements. Because of their complexity and service requirements these systems can only be used in the laboratory. Presently for field applications only compact versions of fixed frequency pulsed UV lasers are appropriate. The wavelengths of several of these laser sources match quite well the absorption spectrum of water contaminated for example by diesel fuel or by gasoline (Fig. 1). These pulsed lasers also allow time resolved spectroscopy for increasing sensitivity [2-4].

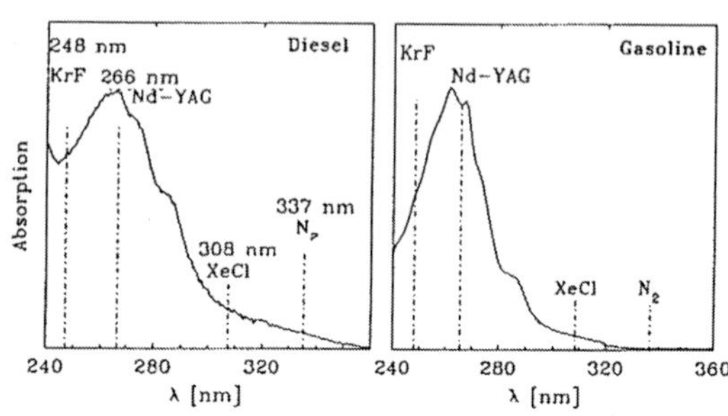

Fig. 1 : Absorption spectra of water contaminated by diesel fuel and gasoline

The frequency upconverted (*4) Nd:YAG laser wavelength at 266 nm coincides with the absorption maximum of several pollutants. Together with a sufficient quantum yield this allows a high detection sensitivity. Very recently, a new solid state laser material (Nd:KGW, λ_0=1067 nm) with high efficiency and low lasing threshold [5,6] has been developed which is very appropriate for use in a mobile laser fluorimeter. For excitation at 248 nm or at 308 nm compact KrF or XeCl excimer lasers [7] can be used. A disadvantage of these lasers in field applications is the need for gas filling changes from time to time.

The transmission behaviour of the optical fibers are crucial for the performance of the laser-fluorimeter. Most appropriate for these applications are multimode fibers with a core of fused silica (high OH content, $\varnothing$ 100µm-1000µm) and a cladding of fluorine doped fused silica [8,9]. Typical fiber

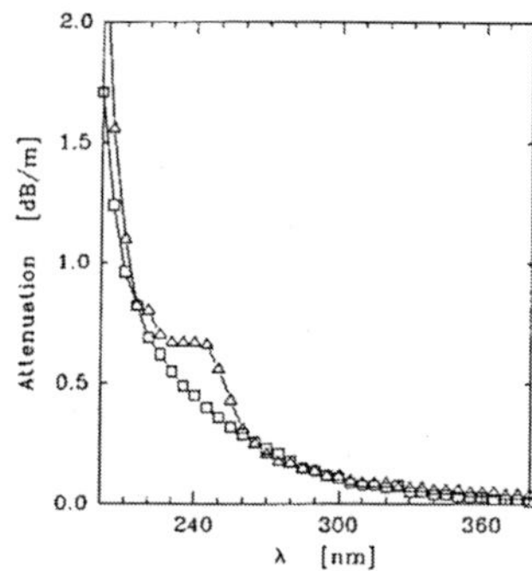

Fig. 2 : Fused silica fiber absorption spectra for two different fibers in the UV range

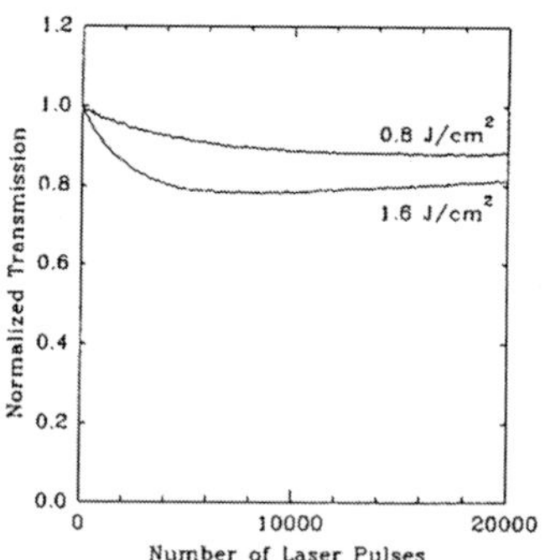

Fig. 3 : Fiber transmission for 308 nm laser radiation at two different coupling energy densities

absorption spectra are shown in Fig. 2. The fiber attenuation is dominated by Rayleigh scattering ($\sim\lambda^{-4}$).

For state of the art fibers the damping coefficients are about 0.15 dB/m at 308 nm and about 0.35 db/m at 266 nm. Some fibers show an additional absorption band around 240 nm. The maximum pulse energy, which can be coupled into the fiber is limited by the surface damage threshold. For 308 nm excimer laser pulses of 30 ns duration we found a typical damage threshold of about 20 J/cm² which decreases for shorter pulses and wavelengths (i.e. 248 nm ,25 ns : 7 J/cm²) [10].

We performed fiber transmission studies over several thousand laser pulses and found that during UV laser transmission the fiber absorption increases. An example is shown in Fig. 3. For 308 nm the photodegradation dT(in dB) of the fiber depends on the laser energy density F (dT~F), the pulse duration τ (dT~τ^{-1}), the laser repetition rate f (dT~(1-exp(-αf))) and on the fiber length l (dT~(1-exp(-βl))) [9,10]. Photodegradation also increases with decreasing wavelength, however the functional form is not yet known. For practical operating conditions with long fibers it seems realistic to take into account laser induced degradation effects which reduce the laser energy at the distal fiber end by 50% of the initial value. For a fiber length of 20 m, an attenuation of 0.35 dB/m (λ=266nm) and induced transmission losses of 50% about 500µJ of laser energy have to be launched into the fiber to achieve an output energy of 40µJ. During irradiation interruptions we observed a (partly) recovery of the fiber transmission.

Spectroscopic Results

We tested the detectability of diesel fuel, engine oil and gasoline (unleaded) with excitation at 248 nm. A laboratory setup with a KrF excimer laser, fiber optics for excitation and detection and an optical multichannel spectrometer with image intensifier was used. Typical spectra obtained in contaminated

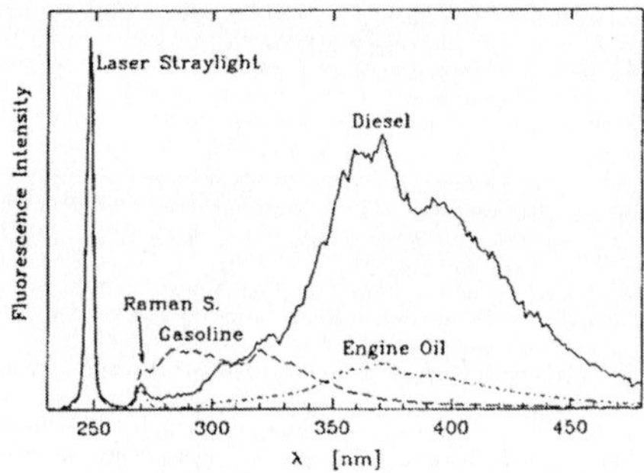

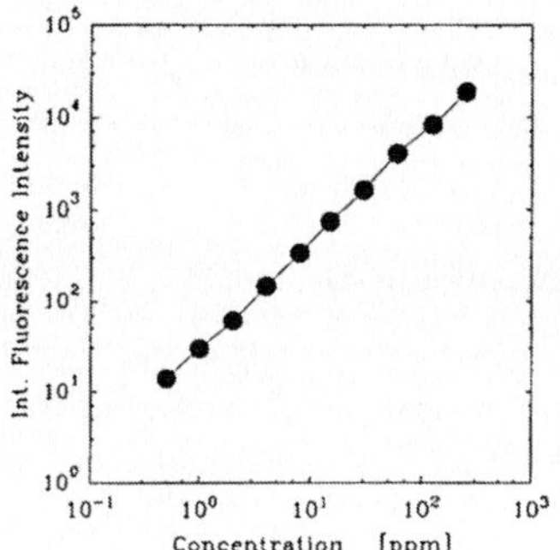

Fig. 4 : Fluorescence spectra of drinking water contaminated by different polluitants. λexc = 248 nm

Fig. 5 : Integrated fluorescence as a function of gasoline contamination

drinking water are shown in Fig. 4. Beside the peak at the laser wavelength (elastic scattering) and around 270 nm (Raman scattering by water molecules) broad fluorescence spectra are observable for all three

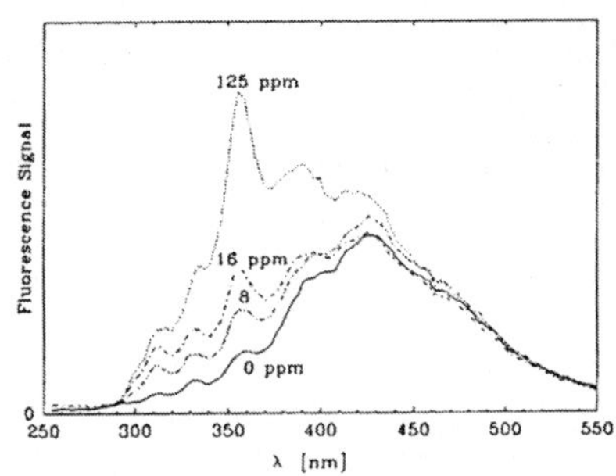

contaminations. Only this spectral information is enough to discriminate between gasoline contamination or pollution by diesel fuel or engine oil. The integrated intensity for different gasoline concentration (Fig. 5) shows that already this preliminary test setup allows a sub-ppm detection limit.

Under certain practical conditions (i.e. for seawater or riverwater) already the natural water contains fluorescent components which produce a background signal. Also the stray light level is increased by particles in the water. In Fig. 6 spectra from riverwater (Ems) contaminated by different amounts of diesel fuel are shown. Laser and Raman scattering was suppressed by a dielectric optical filter. The strong overlapp between pollution fluorescence and background spectrum is obvious. For

Fig. 6 : Fluorescence spectra of river water, contaminated by different concentrations of diesel fuel. λ_{exc} = 248 nm.

quantitative analysis the background signal has to be seperated [11,12] from the signal. Although background effects cause additional fluctuations diesel contamination at concentrations near 1 ppm detection are still detectable. We expect further improvements by signal integration and optimization of the sensor tip design. Time resolved detection will be checked for suppression of laser straylight and background fluorescence [2-4].

The use of the 266 nm wavelength was also tested in the laboratory using a frequency doubled dye laser. For field applications this wavelength will be delivered by a solid state laser. Compared to 248 nm fiber losses are smaller at this wavelength. As an example we studied the detection of phenol solved in water. Phenol, which is known to be toxic, is part of many industrial processes an can be introduced into water accidentally. Detection of phenol at concentration of about 100 ppb is possible with a very good signal to noise ratio (Fig. 7). For this type of contaminations detection limits in the lower ppb range should be realistic.

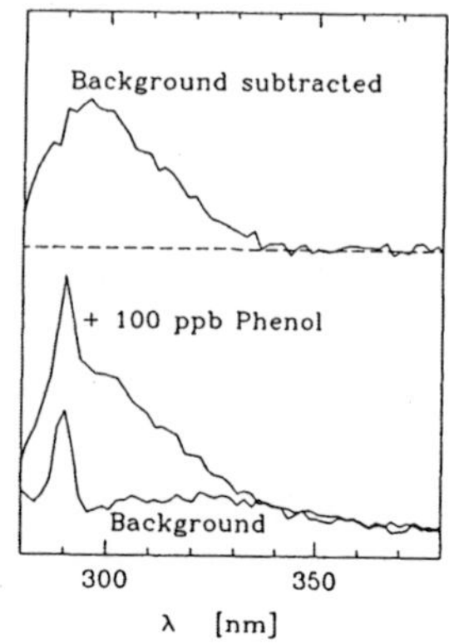

Fig. 7: Fluorescence spectrum of phenol solved in water. λ_{exc} = 266 nm

Conclusions and Outlook

UV laser fluorimeters with fiber optical sensors are well suited for in situ detection of several kinds of water pollutants. Fixed frequency UV lasers can be used in combination with fiber optics. Fiber properties as laser induced attenuation have to be taken into account for fluorimeter applications. In principle the combination of two wavelengths in one device can increase the application range. For example the simultaneous use of the third and the fourth harmonic (355 nm and 266 nm) of a Nd:YAG laser or a combination of 266 nm with 337 nm from a nitrogen laser can be realized in a compact device. New laser materials should also result in an improvement . We will extend our work in these directions.

References

1.   W.A. Chudyk, M.M. Carraba, J.E.Kenny : Anal. Chem. 57, 1237 (1985)
2.   U. Panne, F. Lewitzka, R. Niessner : Analusis 29, 533 (1992)
3.   S.M. Inman, P. Thibado, G.A. Theriault, S.H. Lieberman : Anal. Chim. Acta 239, 45 (1990)
4.   W. Schade, J. Bublitz, V. Helbig, K.P. Nick : in "Laser in remote sensing", eds. C. Werner, V. Klein K.Weber, Springer Verlag
5.   V. Kushawaha, A. Banerjee, L. Major : Appl. Phys. B 36, 239 (1993)
6   K. Stankow, Laser-Laboratorium Göttingen : Private communication
7.   i. e. MPB Technologies Inc., Quebec, Canada, model PSX-100
8.   G. Hillrichs, M. Dressel, H. Hack, R. Kunstmann, W. Neu : Appl. Phys. B. 54, 208 (1992)
9.   H. Fabian, U. Grzesik, G. Hillrichs, W. Neu : appears in SPIE 1893 (1993)
10.   K.F. Klein, G. Hillrichs, W. Neu, H. Fabian, U. Grzesik : appears in "Jahrbuch Laser 1993", Vulkan Verlag
11.   J. M. Andrews, S.H. Lieberman : SPIE 1587, 30 (1991)
12.   B. Abel, H. Hippler, B. Körber, A. Morguet, W. Neu : SPIE 1525, 110 (1991)

A Fiber Optic LIF-Sensor for Measuring Temporal and Spatial Distribution of Tracers in the Ground

A.Knaack[1], D.Klatt[1], W.Schade[1], R.Horn[2], T.Baumgartl[2]

[1]Institut für Experimentalphysik, Universität Kiel

[2]Institut für Pflanzenernährung und Bodenkunde, Universität Kiel

Olshausenstr. 40-60, 24098 Kiel, Germany

Introduction

In soil physics there is a great interest in the exact investigation of the flow path of different solutions in the ground. Both, the spatial and the temporal behaviour should be examined. These values are important sources of data for evaluating the validity of the existing transport models. In environmental techniques these models are used to describe the mobility of different pollutants like plant-protective agents and radio nuclides as well as fertilizers. Present methods, utilizing batch experiments, are of limited value because they only allow to determine breakthrough curves (BTC) using the whole soil sample as a black box. There is no information about the local characteristics in these samples. Usually a soil column is eluted from one side with a dye tracer solution while the effluent is collected in fractions on the other side. The concentrations of the dye in these fractions are determined photometrically [1,2].

The system investigated in our research group is based on the method of laser induced fluorescence (LIF) and uses a system of fiber-optic sensors. This offers the possibility to detect the fluorescence of suitable dye solutions directly in the soil sample. The selection of the tracer, the design of our experiment and the results will be presented in the following paper.

The experimental setup

The schematic design of our experiment is shown in Fig.1a. The emission of an Ar^--laser at 488nm is coupled directly into a fiber optic system. The fiber optic system consists of five excitation and

114

five observation fibers, plus one laser monitoring fiber. This setup
allows the simultaneous use of five optrodes including one for on-line
lasermonitoring. In the optrode the excitation and observation fibers
are fixed together by means of a plastic tube. Since the diameter of a
single fiber is only 200μm, it is possible to obtain an optrode diameter of less than 1mm (Fig.1b). Small optrode diameters minimize the
disruption of the natural structure of the soil sample.

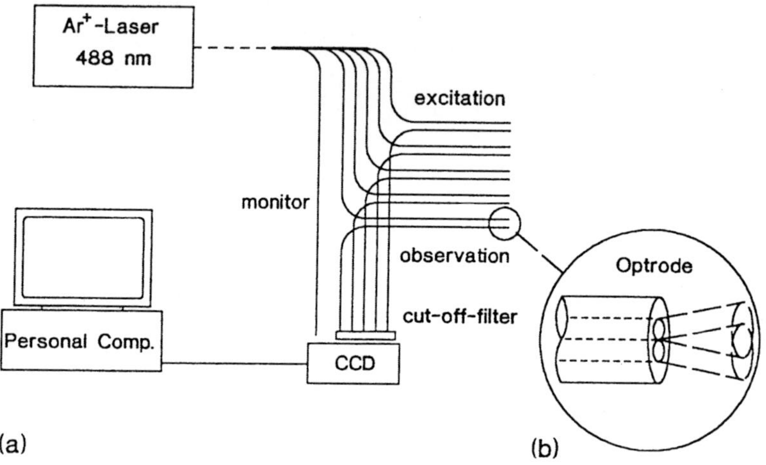

Fig.1:   (a) Schematic setup (b) Optrode

For the detection a CCD-camera is used, which is protected by a cut-
off filter from the backscattered and reflected laser light. The observation fibers are mounted directly on the surface of the filter.
The optrodes are positioned at well defined places in the soil. The
moment the dye solution reaches one optrode the emitted fluorescent
light from this position is received by the CCD-camera and is recorded
by a personal computer that also controls the CCD-camera.

Dye selection

The limiting factors for the selection of the dye solution are given
by the laser wavelength and the requirements of soil physics. Fluorescine is used in this experiment: first because of its suitability regarding the excitation by an Ar$^+$-laser, second, because of its well
known qualities in soil investigations through long standing applications in soil physics. Adsorption of this dye onto sediment surfaces
is not observed for quartz sand but to a certain extend for clay
[3,4]. The major problem, especially for fluorescin, is the dependence
of the fluorescence on the pH-value of the surrounding medium. While

there is no influence on the fluorescence in the alkaline range, a rapid decrease in intensity is observed in acidic media [3]. In the present experiments soil samples are used in which these effects are negligible.

Experimental methods and results

For all experiments quartz sand (which is chemically and physically uncritical for our purpose) with a grain size distribution of less than 2mm was used. This sand was filled in columns of 72mm diameter at heights of 40mm or 60mm. Then the sample was saturated with water from the bottom. Optrodes were positioned as shown in Fig.2. The fluorescin solution (4mg fluorescin/100ml water) was applied to the sample in drops above the position of optrode 1 as it is indicated by an arrow. Directly after application of the dye, water was used in the same manner to rinse the sample.

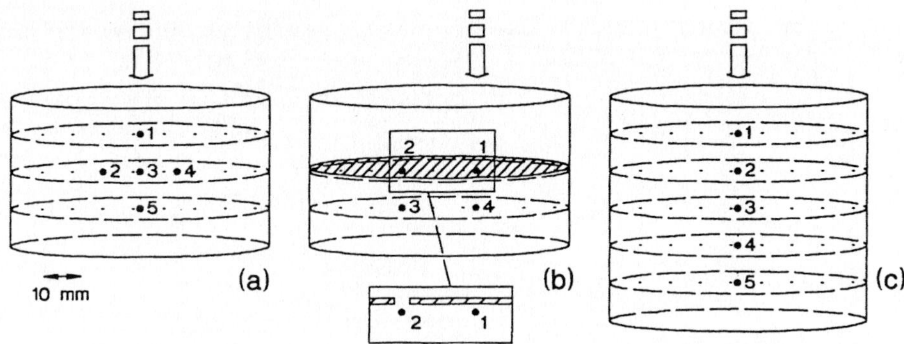

Fig.2: Positions of the optrodes for different experiments

The time dependent increase and decrease of the signals resulting from the variation of the fluorescine concentration at the position of the optrodes in the sample is shown in Fig.3 (refer to Fig.2a). As expected there is a fluorescence signal at position 1 first. Thereafter the signal was observed at position 3 and shortly thereafter, almost simultaneously, at position 2 and 4, and at last at position 5. After 1000 sec water was applied to the sample, and the same principal behavior was observed. To relate a dye concentration to the signal amplitude is not possible, due to factors like soil color or pore volume which may be different at different positions in the sample. But there are some characteristic effects as shown in Fig.4 (refer to Fig. 2c). In this case the same sample was eluted two times using two different drip frequencies.

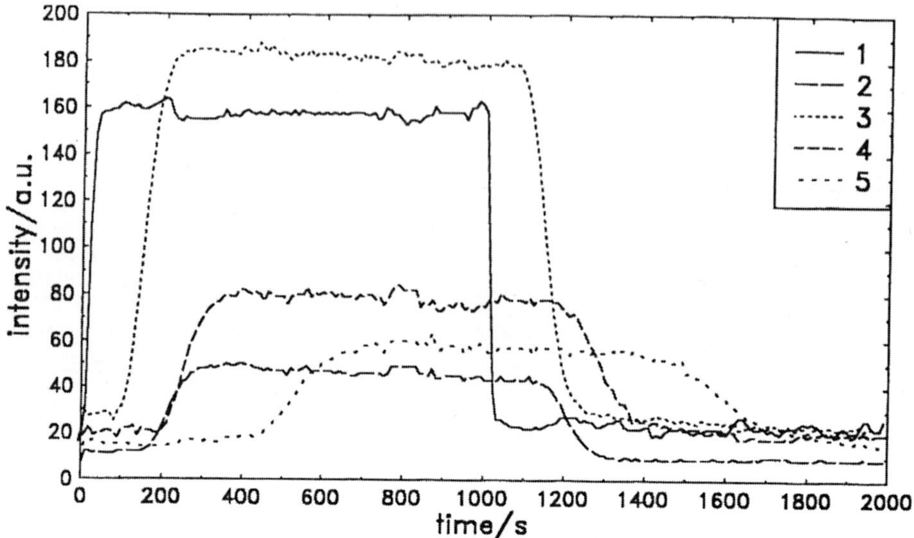

Fig.3: Time dependent fluorescence signals from different positions in the sample

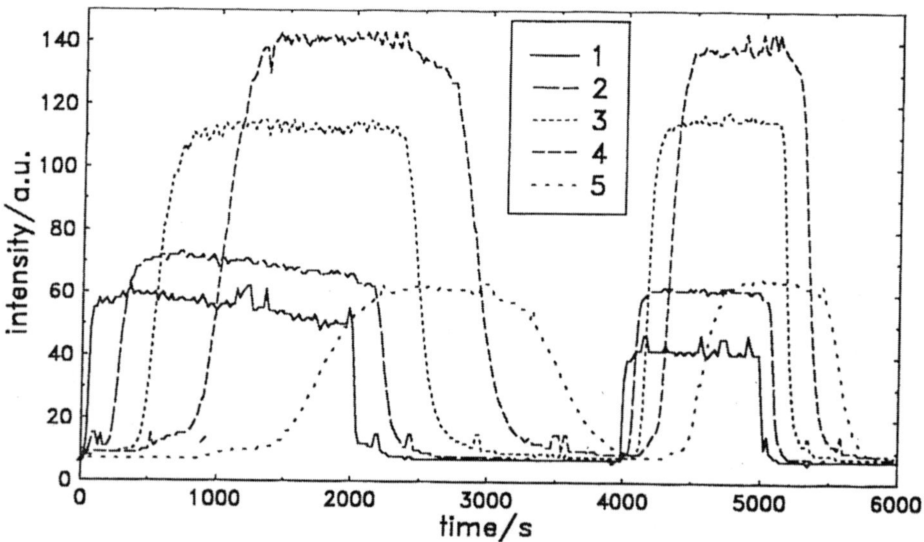

Fig.4: The influence of the drip frequency on the BTC

The BTC obviously depends on the frequency (low fequency for dye until 2000 sec and water until 4000 sec, high frequency for dye until 5000 sec and water to 6000 sec) and also shows an effect caused by the position in the soil. Both, lower frequencies and lower position, give the curves a smoother form. Another series of experiments was engaged with the watermovement through a little hole (1mm diameter) in a light- and watertight plastic-foil (Fig. 2b). This is a simple and idealized model of a defective sealing as it may occur in a landfill. As Fig.5 shows, the dye solution was directed to position 2 and after that there was a hydrodynamic distribution with depth as well as with the position under the foil as the signals at position 1 and 3 show. The dye solution reached position 4 last.

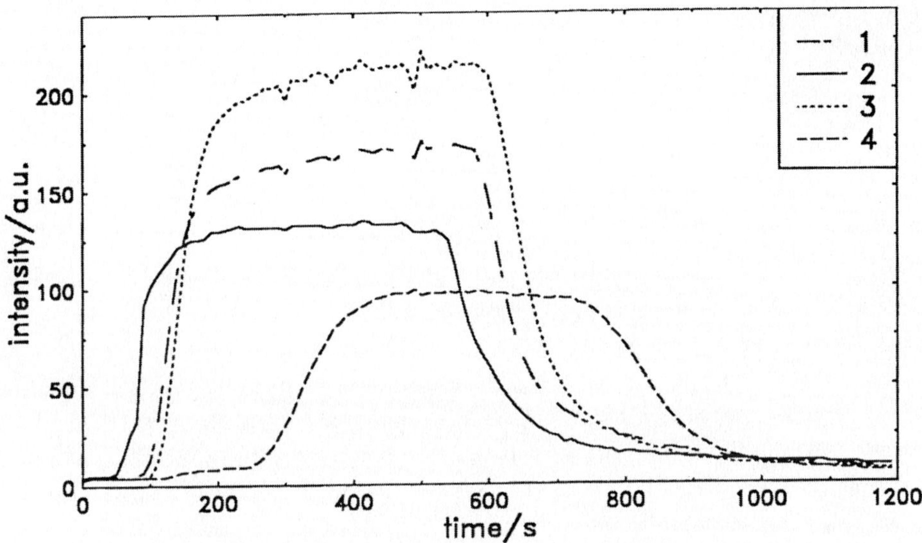

Fig.5: Time dependent distribution of the tracer in a simple model of a landfil with a defective sealing

Conclusion

With the system presented it will be possible to obtain parameters with a resolution in space and time of the distribution of tracers in the ground, which were not measured in soil physics yet.

Although this system was only tested in very idealized soils, there is a good chance to use it in all substrates, which allow the inset of fluorescent dyes. First observations in a clay-sand mixture do indicate this.

References

[1] H.Flühler, R.Schulin, B.Buchter and K.Roth: "Modellierung des Stofftransportes im Boden". Methoden und Konzepte der Bodenphysik, Weiterbildungsseminar der DBG in Kandersteg vom 4.-7. April 1989, Teil B, 1-18 (1990).

[2] S.Koch and H.Flühler: "Solute transport in aggregated porous media: comparing model independent and dependent parameter estimation", accepted for publication for a special issue of "Water, air and soil pollution".

[3] P.L. Smart and I.M.S. Laidlaw: "An evaluation of some fluorescent dyes for water tracing", Water resources research, Vol.13, No.1, 15-33 (1977).

[4] S.Klir: "Untersuchungen über den Intensitätsabfall des Fluoreszeins", Geologisches Jahrbuch, C2 29-33, Hannover 1972

Quantitative Analysis of PAH-Molecules by Time-Resolved LIF-Spectroscopy in Water and in the Ground

J. Bublitz, W. Schade, M. Dickenhausen
Institut für Experimentalphysik der Universität Kiel
Olshausenstraße 40–60, 24098 KIEL, Germany

Introduction

The increasing environmental pollution makes it necessary to develop new techniques for the trace analysis of various pollutants. For the diagnosis of oil pollution on the water surface the method of laser–induced fluorescence spectroscopy has been established during the last years [1,2,3,4]. In comparison to other spectroscopic techniques, like Raman scattering, the measurement of atomic or molecular fluorescence intensities is superior in sensitivity, because of the large cross sections [5]. However, in order to improve the selectivity of this method, not only detecting fluorescence intensities at particular observation wavelengths, but also time–resolved laser–induced measurements have to be applied. Various types of oil products show significant different fluorescence decay spectra between 400 nm and 500 nm after excitation with 337.1 nm according to differing compositions of $\underline{p}$olycyclic $\underline{a}$romatic $\underline{h}$ydrocarbon (PAH) molecules [1,2,3].

In the present work the usually complicated technique for data processing and evaluation of time–resolved measurements has been simplified by reducing the observation of the fluorescence in an *early* and a *late* time–gate in regard to the excitation pulse. The theoretical fundamentals of this method and the capability for a practical application in the environmental trace analysis of the total contermination of PAH–molecules is demonstrated by field measurements in an industrial sewer and by analysing oil–polluted drillings from the ground.

Fundamentals of the method

The following considerations are made for a system of one species of PAH–molecules M in dilute solution with the concentration $[^1\text{M}]$. In case of an excitation with low laser intensity I_L, the molecular fluorescence intensity I_F of the molecules, which should be detected, is proportional to the concentration $[^1\text{M}]$:

$$I_F \propto [^1\text{M}] \cdot I_L \quad . \tag{1}$$

If the excitation is performed by a short laser pulse of negligible duration (δ–function excitation) at $t = 0$, the temporal derivative for $t > 0$ of the concentration of excitited molecules is given by the rate equation [6]:

$$\frac{d\,[^1\text{M}^*]}{dt} = -\left(k_{SE} + k_{IQ} + k_{XM} \cdot [^1\text{X}]\right) [^1\text{M}^*] \quad . \tag{2}$$

The rate coefficients in equation (2) for spontaneous emission k_{SE}, inter- and intramolecular quenching processes k_{IQ} and impurity–quenching k_{XM} caused by an additional species X in the solution are independent of time, provided that the viscosity of the solution is low. In that case the differential equation can be solved by the *fluorescence response function*:

$$I_F(t) = a \cdot e^{-\frac{t}{\tau_F}} \quad , \tag{3}$$

wherein τ_F is the *molecular fluorescence lifetime* of the species M:

$$\tau_F = \frac{1}{k_{SE} + k_{IQ} + k_{XM} \cdot [^1X]} \quad . \tag{4}$$

Since petroleum products are mixtures of a large number of PAH–molecules, the temporal evolution of their fluorescence decay spectra can be described by a sum of exponential functions with characteristic intensities a_i and fluorescence lifetimes τ_i:

$$I_F(t) = \sum_{i=1}^{n} a_i \cdot e^{-\frac{t}{\tau_i}} \quad . \tag{5}$$

Time–resolved laser–induced fluorescence measurements with different oil products like engine oil, diesel fuel or light fuel oil have confirmed this equation [7,8,9]. The fluorescence decay of engine oil (15W–40HD) observed at $\lambda_F = 400$ nm after excitation with a nitrogen laser at $\lambda_L = 337.1$ nm consists of three exponential functions:

$$I_{FO}(t) = a_1 \cdot e^{-\frac{t}{\tau_1}} + a_2 \cdot e^{-\frac{t}{\tau_2}} + a_3 \cdot e^{-\frac{t}{\tau_3}} \quad , \tag{6}$$

with $a_1 = 730$, $\tau_1 = 7$ ns, $a_2 = 240$, $\tau_2 = 27$ ns, $a_3 = 30$ and $\tau_3 = 110$ ns. Beyond that, different media, like water or soil, in which petroleum products will be detected, show also fluorescence spectra with specific relaxation times [7,8,9]. For natural water containing dissolved organic matter (DOM) also called Gelbstoff [5] a fluorescence signal with two decay times of $\tau_4 = 4$ ns, $\tau_5 = 10$ ns and corresponding intensities $a_4 = 650$ and $a_5 = 350$ is observed at $\lambda_F = 400$ nm after excitation at $\lambda_L = 337.1$ nm (refer Fig. 1b(1)):

$$I_{FW}(t) = a_4 \cdot e^{-\frac{t}{\tau_4}} + a_5 \cdot e^{-\frac{t}{\tau_5}} \tag{7}$$

Based on these experimental results, a simple theoretical approach to the change of the fluorescence decay spectra depending on the concentration of dissolved emulsified oil in water has been developed. A mixture of $\frac{x}{1000}$ parts of engine oil and $\frac{1000-x}{1000}$ parts of water causes a variation of the intensity values in the equations (6) and (7):

$$a_{1,2,3} \rightarrow a_{1,2,3} \cdot x \cdot \triangle a \quad \text{and} \quad a_{4,5} \rightarrow a_{4,5} \cdot (1 - x \cdot \triangle a) \quad , \tag{8}$$

where $\triangle a = \frac{1}{1000}$ and $x \in [0, 1000]$. Starting with the fluorescence decay spectra of pure water ($x = 0$) the results of numerical simulations for some increasing values of x are shown in figure 1a.

120

In comparison with the experimental results in figure 1b, measured by a *single–photon–counting*–method for an excitation at $\lambda_L = 337.1$ nm and an observation wavelength of $\lambda_F = 400$ nm, the time–resolved fluorescence signals correspond very well with the results of the theoretical model.

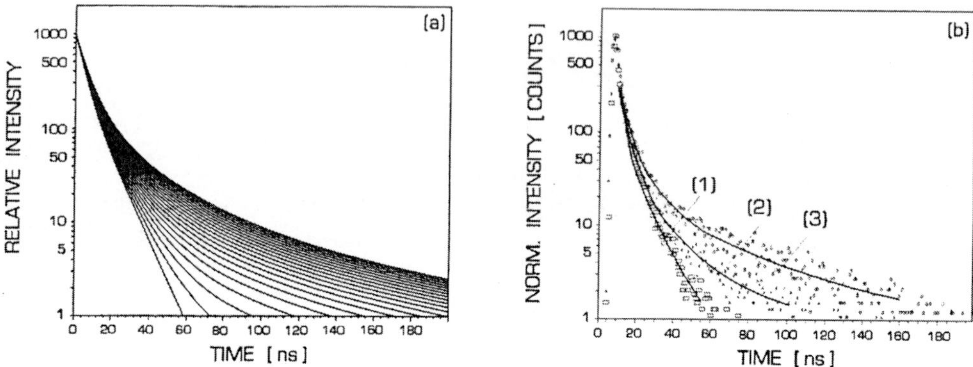

Fig. 1: Time–resolved exponential decay spectra of oil–water mixtures with different concentrations of engine oil (15W–40HD) in Baltic Sea water: a. Results of the numerical simulations, b. Experimental results: (1) pure water, (2) 1 mg/l and (3) 10 mg/l oil in water.

The curves in figure 1 show that engine oil–water–mixtures emit fluorescence even for times longer than 100 ns, while the fluorescence of pure water can only be observed for about 60 ns after the excitation pulse. Therefore the time–resolved detection of the fluorescence was simplified by measuring the fluorescence intensity in an *early* $(0 \leq t \leq t_o)$ and a *late* $(t_o \leq t \leq 2t_o)$ time–interval with respect to the laser pulse. For $t_o = 100$ ns in the first time–window (I_1) mainly the background signal of the water is detected, while the intensity registrated in the *late* interval (I_2) corresponds to the oil fluorescence. An increasing concentration of oil in the water leads to a predominant rise of the fluorescence intensity, measured in the second time–window. So the ratio I_2/I_1 is also mainly determined by the increase of the value of I_2. In the theoretical approach, the ratio of the intensities is equal to a quotient of two linear functions:

$$\frac{I_2}{I_1} = \frac{m_2 \cdot x + n_2}{m_1 \cdot x + n_1} \tag{9}$$

with:

$$m_1 = \triangle a \left[\sum_{i=1}^{3} a_i \, \tau_i \left(1 - e^{\frac{-t_o}{\tau_i}} \right) - \sum_{j=4}^{5} a_j \, \tau_j \left(1 - e^{\frac{-t_o}{\tau_j}} \right) \right] \tag{9a}$$

$$n_1 = \sum_{j=4}^{5} a_j \, \tau_j \left(1 - e^{\frac{-t_o}{\tau_j}} \right) \tag{9b}$$

$$m_2 = \Delta a \left[\sum_{i=1}^{3} a_i\, \tau_j\, e^{\frac{-t_o}{\tau_i}} \left(1 - e^{\frac{-t_o}{\tau_i}} \right) - \sum_{j=4}^{5} a_j\, \tau_j\, e^{\frac{-t_o}{\tau_j}} \left(1 - e^{\frac{-t_o}{\tau_j}} \right) \right] \tag{9c}$$

$$n_2 = \sum_{j=4}^{5} a_j\, \tau_j\, e^{\frac{-t_o}{\tau_j}} \left(1 - e^{\frac{-t_o}{\tau_j}} \right) \quad . \tag{9d}$$

In figure 2a the result for the ratio I_2/I_1 in dependence on x is shown, calculated for the same parameters given by the equations (6), (7) and (8). It is shown, that the quotient I_2/I_1 is a sensitive number for the concentration of PAH–molecules in the oil–water mixtures for PAH–concentrations $[^1M] < 100$ mg/l. This is confirmed by experimental results (refer to Fig. 2b), which have been recieved by an integral detection of the fluorescence intensities of engine oil–water mixtures in various concentrations. The mostly large tolerances of the ratios I_2/I_1 result mainly from an inaccurate determination of the real concentration during the preperation of the emulsions. The present detection limit for a sum contamination of PAH-molecules in water has been determined to 0.5 mg/l by this method.

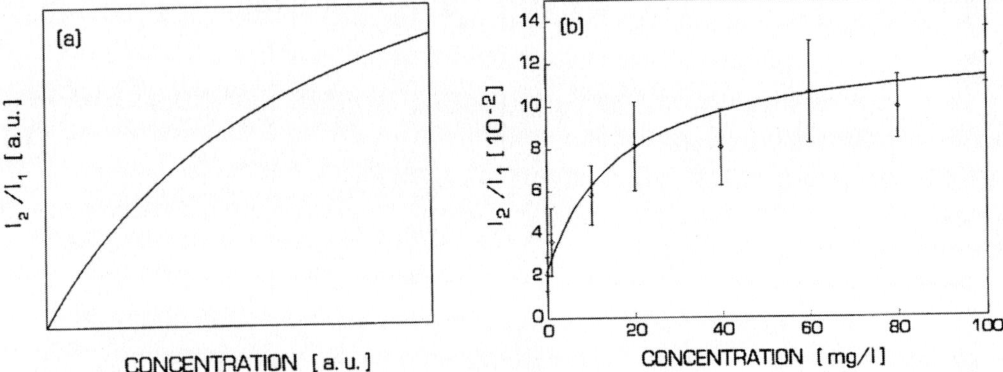

Fig. 2: Ratio of the time–integrated fluorescence intensities I_2/I_1 in dependence on various engine oil (15W–40HD) concentrations in Baltic Sea water: a. Results of the numerical simulations, b. Experimental results.

The experimental set–up for these measurements consists of a compact nitrogen laser with a repetition rate of 10 Hz and a fiber–optic system of separated quartz fibers for excitation and observation of the fluorescence. The emitted photons are detected with a monochromator for wavelength selection ($\lambda_F = 400$ nm), a photomultiplier and two photon counters. The gates of the counters are controled by a trigger unit, which itself is triggerd by a part of the laser pulse. The first gate pulse activates one counter for the time 0–100 ns, while the other counter is started by the second gate pulse for the interval 100–200 ns with respect to the excitation pulse of the laser. The number of photons accumulated by the two counters are transfered and stored by a personal computer after a measuring sequence of one minute.

With this equipment, investigations concerning the applicability of the integrated time–resolved laser-induced fluorescence method for the trace analysis of oil pollution in the ground were done as well.

In figure 3 the results for the ratio I_2/I_1 depending on the concentration of engine oil (15W–40HD) distributed in sand are shown. The integral intensity ratio is qualitatively the same as for oil–water mixtures. Therefore this method is also applicable to the quantitative trace analysis of sum conterminations of PAH–molecules in the ground. Concentrations down to 0.5 mg/kg soil have been detected in laboratory experiments so far.

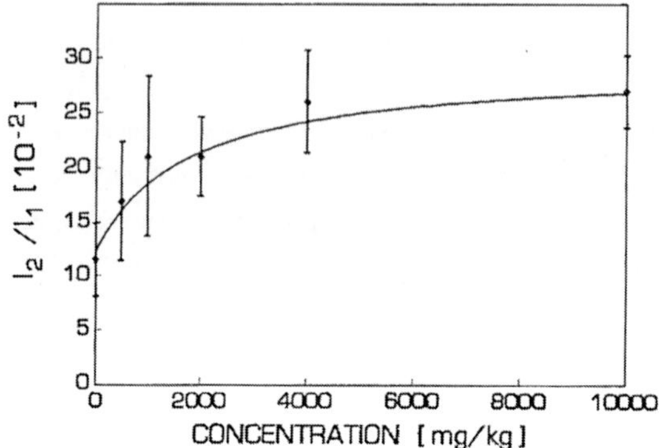

Fig. 3: Ratio of the time–integrated fluorescence intensities I_2/I_1 for various engine oil concentrations in sand.

Results of field measurements

In the first example the capability of the described method for environmental trace analysis of oil–pollution in water is demonstrated by field measurements in an industrial sewer.

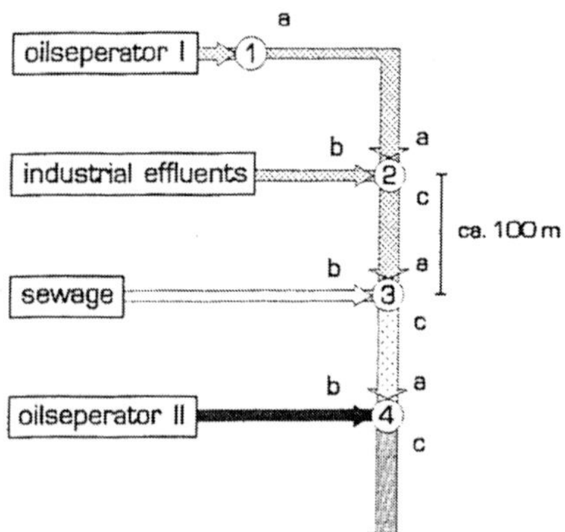

Fig. 4: Scheme of the industrial sewer

	I_2/I_1 $[10^{-2}]$			Concentration $[mg/l]$		
	a	b	c	a	b	c
1	3.1	—	—	1	—	—
2	3.1	3.8	3.9	1	4	5
3	3.9	2.5	3.4	5	<1	2
4	1.3	9.6	3.0	<0.5	60	1

Tab. 1: Measured intensity ratios and the corresponding concentrations

These measurements have been performed with the same experimental set–up and wavelengths (λ_L = 337.1 nm, λ_F = 400 nm) mentioned before. The data were taken at ten different points, which are shown in figure 4. The ratios I_2/I_1 were detected on–line for each point three times by averaging the fluorescence intensities in the two time intervals for one minute. On the basis of the experimental data for engine oil (15W–40HD) given in figure 2b these relative values I_2/I_1 were transformed into concentrations as shown in table 1.

Up to point 3a a sum contamination of PAH–molecules in the range between 1 mg/l and 5 mg/l is observed. These numbers are in good agreement with an average oil concentration of 5 mg/l determined by a chemical analysis, based on the infrared spectroscopic method 'DIN 38409-H18'. Also remarkable is the detected dilution effect by the water of the sewage at point 3c and after a distance of about 100 m at

point 4a. Another essential result of this field experiment is, that the time–integrated fluorescence measurement offers the possibility of detecting a short–term increase of the oil concentration up to 60 mg/l at point 4b, because of the on–line technique. Large fluctuations of the oil content in the water coming from the oilseperator II, which have been observed, could not definitely be detected by spot–checks with a chemical analysis of a sample of the contaminated water.

In the second example an application of the method for the trace analysis of petroleum products in the ground is shown. Drillings, 1 m in length, taken from the ground under a petrol station were analyzed immediately after sampling. In intervals of 50 cm beginning at the top of the drilling, the fluorescence intensity was measured in the two timegates for one minute under the same conditions as in the previous experiments. A site plan of the petrol station, containing the positions of the six drillholes as well as the results of the integral fluorescence detection, are shown in figure 5.

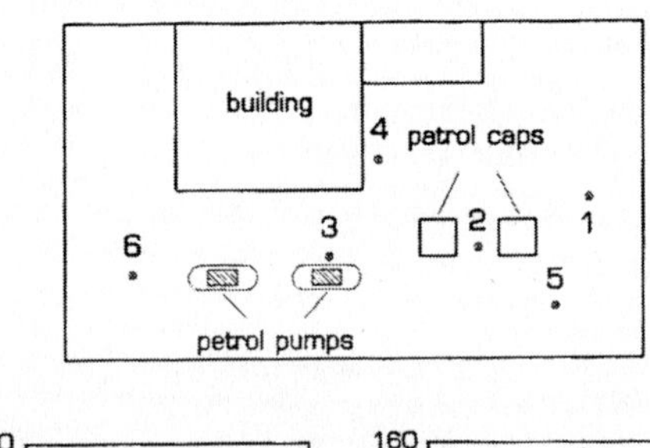

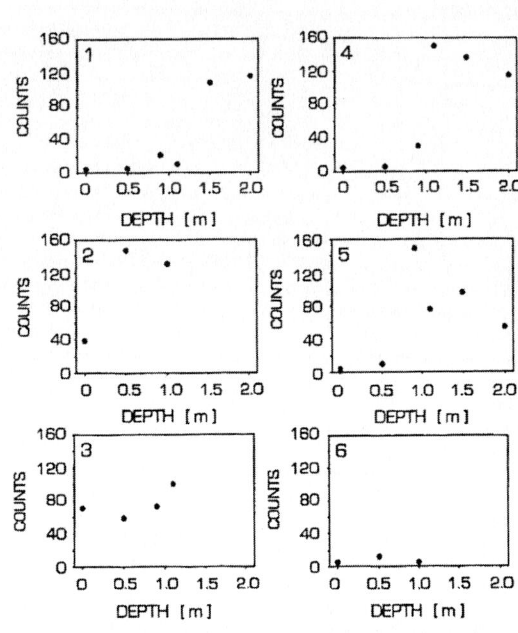

Fig. 5: a. Site plan of the patrol station with the positions of the drill–holes.
b. Results for the integral fluorescence intensities $I_1 + I_2$ in dependence of the drilling depth.

Here the total fluorescence intensity $(I_1 + I_2)$ is taken as an indicator for the relative oil concentration in the ground. In the area of the petrol caps (point 2) and at the petrol pumps (point 3) an increase in the fluorescence intensity is already detected at the surface and also down to a depth of 1 m. These measurements are in good agreement with the results of the chemical analysis done for these samples, which gives sum concentrations of PAH–molecules up to 5900 mg/kg soil. At the drill–holes 1,4 and 5 an oil–contamination is also measured for samples taken below 1 m from the surface. This might be an indication of a distribution of the benzine–polluted groundwater starting at a level of about 1.7 m. The measurements at point 6 yield to an increased fluorescence signal only in a depth of 0.5 m from the top. This is also confirmed by the laboratory chemical analysis, which shows a concentration below 20 mg/kg soil for the surface and the depth of 1 m, while for samples taken below 0.5 m the sum contamination of PAH–molecules is 370 mg/kg soil. The results of these field measurements demonstrate, that this new method of the integral time–resolved laser–induced fluorescence spectroscopy is also suitable for the trace analysis of oil pollutants in the ground. The advantage of this technique compared to the chemical methods is the on–line diagnosis of the samples and an improved spatial resolution in combination with a high sensitivity.

References

[1] R.M. Measures, W.R. Houston, D.G. Stephenson, Opt. Eng., **13** (6), (1974), 494–501

[2] D.M. Rayner, A.G. Szabo. Appl. Opt., **17** (19), (1978), 1624–1630

[3] P. Camagni, G. Colombo, C. Koechler, A. Pedrini, N. Omenetto, G. Rossi, IEEE Transactions on Geoscience and Remote Sensing, **GE–26** (1), (1988), 22–26

[4] T. Hengstermann, R. Reuter, Appl. Opt., **29** (22), (1990), 3218–3227

[5] R.M. Measures, *Chem. Analysis* **50**, *Analytical Laser Spectroscopy*, ed. N. Omenetto, Wiley, New York, (1979), 295 ff

[6] J.B. Birks: *Photophysics of Aromatic Molecules*, John Wiley & Sons Ltd., London, (1972)

[7] W. Schade, J. Bublitz, V. Helbig, K.-P. Nick, *Laser in der Umweltmeßtechnik*, ed. V. Klein, K. Weber, Ch.P. Werner, Springer–Verlag, Berlin, (1992), 53 ff

[8] W. Schade, J. Bublitz, Inst. Phys. Conf. Ser. **128**: Section 9, Inst. Phys Pub., Bristol (1992)

[9] J. Bublitz: diploma thesis, Kiel, (1992)

Acknowledgements

This work was financially supported by Atlas Elektronik GmbH in Bremen, the Gesellschaft für Umwelt– und Sanierungstechnologie mbH (GSU) in Flensburg and the BMFT.

DLR-LIDAR-Activities

Flugzeug-Lidar-Systeme im Einsatz bei der DLR - Eine Übersicht über Methoden und Ergebnisse

W. Renger
DLR-Institut für Physik der Atmosphäre, Münchner Str. 20, D - 82230 Wessling

Einleitung: *LIDAR das optische Pendant zum akustischen Echo*

LIDAR- wie Radarverfahren beruhen in ihrer Grundkonzeption auf dem seit langem in der Akustik bekannten Echophänomen. Ein von einem Sender ausgesandtes Wellenpaket wird an Hindernissen reflektiert bzw. gestreut. Aus der Laufzeit zwischen Aussenden und der Ankunft des Echos am Standort des Senders läßt sich die Entfernung zwischen Sender und Hindernis bestimmen. Beim LIDAR liegt die Wellenlänge typisch im Bereich 0.3-10 µm. Während bei Radargeräten entsprechend ihrer Wellenlänge im cm - und dm - Bereich der Einfluß der Atmosphäre, abgesehen von dichten Wolken, vernachlässigbar ist, untersucht man mit LIDAR-Geräten gerade diese Wechselwirkung, um auf den Zustand bzw. die Zusammensetzung der Atmosphäre zu schließen.

Die Idee, künstliche Lichtquellen für Untersuchungen der Atmosphäre einzusetzen, geht schon auf den Beginn dieses Jahrhunderts zurück. Mit der Erfindung des Lasers und der Entwicklung der ersten Impulslaser zu Anfang der sechziger Jahre standen Lichtquellen zur Verfügung, die es erlaubten aufgrund ihrer Schmalbandigkeit, ihrer bis dahin unerreicht geringen Divergenz und der hohen Impulsleistung von ca. 100 MW bei Impulsdauern von 10-20 nsec, die Meßqualität ganz erheblich zu steigern. Ab 1965 wurden bei der DLR, der damaligen DVL in München-Riem, LIDAR-Systeme entwickelt und für Forschungszwecke eingesetzt. Mit einem Rubinlasersystem konnte 1967 die sogenannte Jungeschicht - ein Gebiet erhöhter Konzentration von stratosphärischem Aerosol nachgewiesen werden. [KATÈRLE, 1969]

Diese ersten Lidarsysteme basierten auf nicht abstimmbaren Festkörperlasern, zunächst dem Rubinlaser. Oberstes Kriterium für die Laserauswahl war die Erzeugung möglichst hoher Impulsenergien bei kurzen Impulsdauern, um das Signal/Rauschverhältnis und die Entfernungsauflösung verbessern zu können. Weitere Voraussetzung war eine Wellenlänge innerhalb der atmosphärischen Fenster und in Bereichen , in denen empfindliche Empfänger zur Verfügung standen. Die Entwicklung des Nd:YAG - Lasers bedeutete einen Sprung nach oben für LIDAR-Anwendungen. Schußfolgen von bis zu 20 Hz wurden möglich. Die Stabilität und die Zuverlässigkeit wurde aufgrund der im Vergleich zum Rubin geringeren thermischen Empfindlichkeit zusammen mit dem höheren Konversionswirkungsgrad erheblich verbessert. Auch die Verluste bei einer Frequenzverdopplung zur Erzeugung der 2. Harmonischen wurde durch die bei 532 nm viel empfindlicheren Multiplier bei weitem aufgehoben.

Die Entwicklung eines Raman - LIDARs erschien wegen der Möglichkeit, die Konzentration unterschiedlicher Gase relativ zur in der Regel bekannten Stickstoffkonzentration zu bestimmen, bestechend. Sie wurde aber wegen der kleinen Raman-Streuquerschnitte und der dadurch nur bedingten Tauglichkeit für Fernmessungen nicht verfolgt.

Mit dem Erscheinen abstimmbarer Laser, zunächst blitzlampengepumpte Farbstofflaser, kamen zu Beginn der siebziger Jahre Resonanzstreu - Lidarsysteme zum Einsatz, die hier nur der Vollständigkeit halber erwähnt werden. Sie dienten in erster Linie zur Erforschung der Natriumschicht in 85 -100 km Höhe. Das Institut für Physik der Atmosphäre hatte damals seine Forschungsarbeiten auf die Troposphäre beschränkt.

Gleichzeitig erschlossen diese Laser auch Bereiche, in denen die in der Atmosphäre vorhandenen Gase die für ihre chemische Zusammensetzung typischen Absorptionslinien besitzen. Durch Messung auf der Absorptionslinie und möglichst dicht daneben ist es auf sehr elegante Weise möglich, die Konzentration des absorbierenden Gases entfernungsaufgelöst zu bestimmen. Der folgende Vortrag wird sich mit einem derartigen System befassen.

Betrachtet man das Signal eines Lidarsystems spektral hoch aufgelöst, so lassen sich zusätzliche Informationen gewinnen. Bewegt sich das streuende Medium relativ zum Lidarsystem, so erfährt das zurückgestreute Licht eine Dopplerverschiebung, aus der sich die Projektion des Geschwindigkeitsvektors auf die Laserschußrichtung ermitteln läßt. Ein weiterer Vortrag wird sich mit dieser Zielsetzung befassen. Über die unterschiedliche Geschwindigkeitsverteilung des atmosphärischen Aerosols und der Moleküle läßt sich der Molekül- vom Aerosolstreuanteil trennen. Aus der Breite des Molekülanteils kann die Temperatur abgeleitet werden.

Mobiles LIDAR

Es dauerte in der Regel Tage, ein System einsatzbereit zu machen und dann hieß es auf wolkenfreien Himmel zu warten. Spielte das Wetter und die Technik mit, so wäre oft ein rascher Wechsel des Standorts dringend geboten. Daraus entstand bald der Wunsch nach einem mobilen System. Allerdings erforderten diese ersten LIDAR-Systeme erhebliche Anschlußleistungen und eine entsprechende Kühlung. Sie waren schwer und unbeweglich und auch die Datenerfassung mittels Oszillographen und Polaroidphotos war sehr unhandlich. Dementsprechend wurde das erste "mobile System" der DLR auf einem Forschungsschiff eingesetzt.

Ungeachtet dessen entstanden aber bereits 1968 die ersten Konzepte mit dem LIDAR in die Luft und noch höher, auf eine Erdumlaufbahn zu gehen. Es folgten Studien in Zusammenarbeit mit der ESRO, der Vorgängerorganisation der ESA, und der NASA für den Einsatz eines Rückstreu-LIDARs beim ersten Spacelabflug, die dann leider nicht verwirklicht wurden. [FIOCCO et al. 1976]

LIDAR im Flugzeug

Parallel dazu wurden Pläne für ein Flugzeugsystem entwickelt. Ein Flugzeug war bzw. ist ein idealer Träger, da nahezu jeder beliebige Punkt der Erde erreichbar ist. Auch völlig unzugängliche Gebiete können untersucht werden. Aufgrund der hohen Geschwindigkeit sind "Momentaufnahmen" möglich. Durch die zwar nur in Grenzen freie Wahl der Flughöhe kann der Abstand LIDAR - Meßobjekt optimiert werden.

Ein Problem bodengebundener Systeme, die in der Regel senkrecht nach oben sondieren, liegt in der hohen Dynamik des Meßsignals, die verarbeitet werden muß. Das Meßsignal nimmt rein geometrisch - optisch mit dem Quadrat der Entfernung ab. Die Luftdichte halbiert sich alle 5,2km. Der Volumenrückstreukoeffizient der Atmosphäre, zu dem das Meßsignal proportional ist, nimmt aber entsprechend der Abnahme der Aerosolteilchenkonzentration noch viel rascher ab und damit auch das Nutzsignal. Bei einem von oben nach unten sondierenden System sind diese Eigenschaften gegenläufig. Hierdurch sind prinzipiell höhere Meßgenauigkeiten erzielbar.

LIDAR-Signale sind aus technischen Gründen erst nach einer Mindestentfernung von typisch 300 -1000 m zuverlässig auswertbar. Ist der Transmissionsgrad auf dieser Strecke deutlich kleiner 1, so läßt sich die Extinktion des Laserlichts auf dieser Strecke nur schlecht abschätzen. Dies ist besonders dann sehr ungünstig, wenn sich die optischen Eigenschaften der Atmosphäre rasch ändern. Dies ist für bodennahe Messungen häufig der Fall, bei Flugzeugmessungen praktisch nie.

Bei der DLR waren Forschungsflugzeuge vorhanden, die bereits über eine für den Betrieb von wissenschaftlichen Geräten erforderliche Grundausrüstung verfügten. Deshalb war die Planung für ein Flugzeuglidar nicht nur wünschenswert, sondern auch eine wesentliche Voraussetzung für die Verwirklichung gegeben. Doch ganz überraschend mit sehr engen Terminvorgaben ergab sich zunächst eine Mitfluggelegenheit auf einem Flugzeug der NASA.

ASSESS

Im Rahmen der Spacelab/Shuttle Vorbereitungen führte die NASA einen Simulationsflug mit einem Düsenverkehrsflugzeug durch, bei dem die äußeren Bedingungen eines Spacelabflugs mit Ausnahme der Schwerelosigkeit erprobt werden sollten. Für dieses Programm wurde von der DLR ein kleines Rückstreulidar zur Verfügung gestellt. [WERNER, 1978]

ALEX-F in der Falcon

Durch die Beschaffung eines Flugzeuges vom Typ Falcon 20 als meteorologischen Forschungsflugzeuges der DLR, stand ab Mitte der siebziger Jahre ein geeignetes Flugzeug für komplexe Geräte zur Verfügung. Dieses Flugzeug hatte - erstmalig bei der DLR - ein Trägheitsnavigationssystem, zwei große optische Fenster von 50 cm Durchmesser im Boden, ausreichende elektrische Stromversorgung und eine Druckkabine, die den Betrieb von Meßinstrumenten unter quasi Laborbedingungen ermöglichte. Nach einer Vorversion, die noch einen leistungsschwachen Nd:Glass Laser verwendete, konnte das Bundesministerium für Forschung und Technologie überzeugt werden, ein Flugzeug-LIDAR als Vorstufe für eine Anwendung im Raum zu unterstützen. Das Aerosollidarexperiment ALEX F wurde in Zusammenarbeit mit der Firma Dornier System entwickelt und wird in wesentlichen Komponenten noch bis heute eingesetzt. Nur die digitale Datenerfassung und Speicherung wurde in den letzten Jahren von Grund auf erneuert. Das System war in einer Vielzahl von Kampagnen von der Arktis bis Nordafrika für Aufgaben des Umweltschutzes und im Rahmen der Fernerkundung meteorologischer Parameter eingesetzt. Es wurde etwa der Transport von Luftverschmutzungen aus Kraftwerksanlagen, der Transport und die Verteilung des Aerosols in Gebirgstäler, in Industriegebieten wie das Raffineriegebiet um Ingolstadt aber auch um Fos-Berre in Südfrankreich untersucht. Die Ausbildung der Grenzschicht und ihr Tagesgangs wurde ebenso studiert wie das Auftreten des "Arctic Haze" oder die Ausbreitung künstlich erzeugten Nebels über der Nordsee. [MÖRL et al. 1981]

ALEX-F in der Do 228

Immer wieder gab es Problemstellungen, bei denen man gerne vom Flugzeug aus nach oben sondiert hätte. Bei der Falcon war aus Kostengründen der Einbau eines entsprechenden Fensters nahezu ausgeschlossen. Bei der Planung für die Ausstattung einer DO 228 konnte dann aber erreicht werden, daß zwei in einer Achse liegende Öffnungen an der Ober- und Unterseite vorgesehen wurden. Durch Anordnung des Teleskops in der Flugzeuglängsachse und einen vorgesetzten Schwenkspiegel unter 45° war es möglich vertikal nach oben wie nach unten zu sondieren. Der Wechsel von Zenit- auf Nadirsondierung läßt sich in wenigen Sekunden auch im Flug bewerkstelligen. Gerade wegen dieser Flexibilität, aber auch wegen der oftmals besseren Verfügbarkeit der Do 228, hat sich diese Version gut bewährt. Nachteilig ist allerdings die hohe Vibrationsbelastung der Geräte und die fehlende Druckkabine, wodurch die maximale Flughöhe auf gut 3000 m begrenzt wird. 1989 wurde mit diesem System erstmals eine Leewellen-PSC in Nordskandinavien vermessen. Im Oktober 92, während der Cirruskampagne 92, könnten Rückstreuproilfe bis über 20 km Höhe ausgewertet und die optische Dicke des Pinatubo-Aerosols in diesem Höhenbereich abgeschätzt werden.

Wasserdampf-DIAL in der Falcon

Nach den ersten erfolgreichen Flugzeug - Lidarkampagnen begann die Entwicklung eines abstimmbaren Lidarsystems für die Bestimmung der Wasserdampfkonzentration in der freien Atmosphäre. Ziel war eine Vertikalauflösung von 100 bis 200 m. Die Horizontalauflösung sollte < 500 m sein. Auf der Basis eines Nd:YAG gepumpten Farbstofflasers wurde das System zur Einsatzreife entwickelt. Eine ausführlichere Darstellung bringt der folgende Vortrag.

Ozon-LIDAR-Experiment auf der Transall OLEX

Mit der zunehmenden Sorge über einen möglichen Ozonabbau in der arktischen Stratosphäre, wie er in der Antarktis seit einigen Jahren in bedrohlichem Umfang gemessen wurde, begannen im Jahre 1989 die Planungen im Rahmen des Ozonforschungsprogramms für den Einsatz einer Transall der Bundeswehr. Bei der experimentellen Ausstattung der Maschine wurde auch der Vorschlag des Instituts für Physik der Atmosphäre für ein LIDAR berücksichtigt. Das bestehende Rückstreulidarsystem wurde für Ozon-DIAL-Messungen um einen Verdreifacher für die Referenzwellenlänge bei 354 nm und einen XeCl-Laser (308 nm) der Firma Sopra erweitert sowie die Empfangsseite entsprechend modifiziert. Die Integration in die Transall erfolgte in enger Zusammenarbeit mit der Firma Aerodata, Braunschweig. Ein weiterer Vortrag dieser Reihe befaßt sich mit den sehr erfolgreichen Ergebnissen dieses Systems und ihrer Bedeutung auch für die anderen an Bord der Transall befindlichen Meßgeräte.

Ausblick

WIND-Lidarsystem auf der Falcon

In deutsch-französischer Zusammenarbeit wird an der Entwicklung eines für den Einsatz im meteorologischen Forschungsflugzeug ‚Falcon 20, der DLR gearbeitet. Dieses auf der Basis eines gepulsten CO2- Lasers geplante System wird in einem weiteren Vortrag vorgestellt.

LIDAR auf dem hochfliegenden Forschungsflugzeug Strato 2C

Für den Einsatz in diesem Flugzeug, das eine maximale Reichweite von 18 000 km und eine maximale Flughöhe von 26 km haben soll, wird mit dem System Adler eine neue Generation LIDAR vorgeschlagen. Neu soll vor allem der Konversionswirkungsgrad, die Zuverlässigkeit, der zugängliche Wellenlängenbereich und der Automatisierungsgrad sein. Ziel ist die Entwicklung der Einsatz eines mit Dioden gepumpten Nd:YAG-Lasers, der seinerseits einen abstimmbaren Ti:Saphir pumpt oder über einen optischen parametrischen Oszillator einen großen Wellenlängenbereich zugänglich macht. Im Endausbau soll das System nach entsprechender Einweisung von Experimentatoren (evtl. Piloten) ohne besondere LIDAR-Kenntnisse bzw. Ausbildung betrieben werden können.

LIDAR im Weltraum

Mehr als 20 Jahre wird bei der NASA und der ESA der Einsatz eines Weltraumlidars geplant. Die DLR war über die Jahre immer entscheidend an dieser Planung beteiligt. Vorschläge für die D2 Mission, für Eureca, wurden in Zusammenarbeit mit dem Meteorologischen Institut in München, dem Batelle-Institut in Frankfurt und der Firma Dornier entwickelt. [MÖRL et al.] Ein wesentlicher Grund, daß die Planungen bisher nicht umgesetzt werden konnten, lag in den zunächst immer zu großzügig angesetzten Energieressourcen der Trägersysteme einerseits und dem hohen Srombedarf der geplanten LIDAR-Experimente. Im nächsten Jahr wird LITE mit dem Shuttle für 7 bis 10 Tage in eine Erdumlaufbahn kommen. Noch kurz vorher soll ALISSA, eine russisch französisches Experiment auf einer russischen Raumstation fliegen. Beide Experimente sind auf kurze Lebensdauer ausgelegt und verwenden noch konventionelle mit Blitzlampen gepumpte Nd:YAG - Laser. Die heute vor der Tür stehenden diodengepumpten Systeme stellen den größten Schritt der letzten Jahre in Richtung auf ein operationelles LIDAR im Weltraum dar.

Literatur:

Katerle, H. J.: Die Messung der Rückwärtsstreuung an Molekülen und Aerosolen in der Stratosphäre, Dissertation, LMU München , (1969)

Fiocco, G., Chanin, M.L, Renger, W., Thomas, L.: Spacelab Borne Lidar, Report on the Phase A Study, Volume 1, Scientific Objectives, ESA Rep. DP/PS (76) 4 (1976), 19 S.

Werner, C., Airboprne Lidar Aerosol Measurements over U.S. - Results of the ASSESS II Spacelab Simulation Mission, 4th Symposium on Met. Observations and Instrumentation , AMS, Boston, Mass, (1978)

Moerl, P., Reinhardt, M.E., Renger, W., Schellhase, R.: The Use of the Airborne LIDAR-System "ALEX-F1" for Aerosol Tracing in the Lower Troposphere, Beitr. Phys. Atmosph. Vol 54, S. 403-410 (1981)

Mörl, P., Quenzel, H., Renger, W., Schmitz-Peiffer, A., Schumann, U., Weber, E., Wieg-ner, M.: ALEXIS Atmospheric LIDAR Experiment in Space, Phase A Study DFVLR-Mitt. 88-10, 84 S (1988)

Untersuchung von Wasserdampfverteilungen in der freien Atmosphäre mit einem flugzeuggetragenen Differential-Absorptions-Lidar (DIAL)

G. Ehret, Ch. Kiemle, W. Renger, G. Simmet
Institut für Physik der Atmosphäre, DLR
D-82230 Oberpfaffenhofen

Einleitung

Der Wasserdampfgehalt der Atmosphäre ist ein wichtiger meteorologischer Parameter, der das Wetter- und Klimageschehen auf der Erde nachhaltig beeinflußt. Einerseits verhält sich das Wassermolekül wegen den starken Absorptionsbanden im Infraroten wie ein typisches Teibhausgas und andererseits findet wegen Kondensation und Wolkenbildung der Luftfeuchte eine starke Dämpfung der Sonneneinstrahlung statt. Beides hat in mehr oder weniger komplexer Weise Einfluß auf den Treibhauseffekt. Die Kenntnis von großräumigen Wasserdampfverteilungen in der Atmosphäre und deren zeitliche Veränderungen sind für dessen Verständnis von grundlegender Bedeutung.

Es ist bekannt, daß atmosphärische Feuchtefelder wegen der zahlreichen Quellen und Senken und der verschiedenen Durchmischungsvorgängen in der Atmosphäre zeitlich und räumlich sehr variabel sind. Dies hat zufolge, daß der Wasserdampf mit konventionellen "In situ"-Meßmethoden nur lückenhaft ermittelt werden kann. Mit einem gepulsten Laser können solche Feuchteverteilungen in Ausbreitungsrichtung der Laserstrahlung von einem festen Ort aus "remote" gemessen werden. Werden solche Messungen z.B. vom Flugzeug aus durchgeführt, wobei der Laser entweder nach oben oder unten gerichtet ist, lassen sich diese atmosphärischen Feuchteverteilungen sowohl in vertikaler als auch in horizontaler Richtung untersuchen. Solche Messungen sind notwendig, wenn die Verdunstung über einem vorgegebenen Areal bestimmt oder der Wasserdampftransport bzw. der Wasserdampffluß in der Atmosphäre großräumig untersucht werden soll.

Das DIAL-Verfahren

Bei der Bestimmung von Wasserdampfverteilungen in der Atmosphäre mit einem Laser hat sich das Differential-Absorptions-Lidar (DIAL)-Verfahren, angewandt im nahen Infraroten, bewährt. Dieses Verfahren beruht auf der Absorptionsspektroskopie an gasförmigen Substanzen unter Atmosphärenbedingungen. Im Unterschied zu einem einfachen Rückstreulidar, daß mit Laserpulsen einer Wellenlänge auskommt, messen DIAL-Systeme die

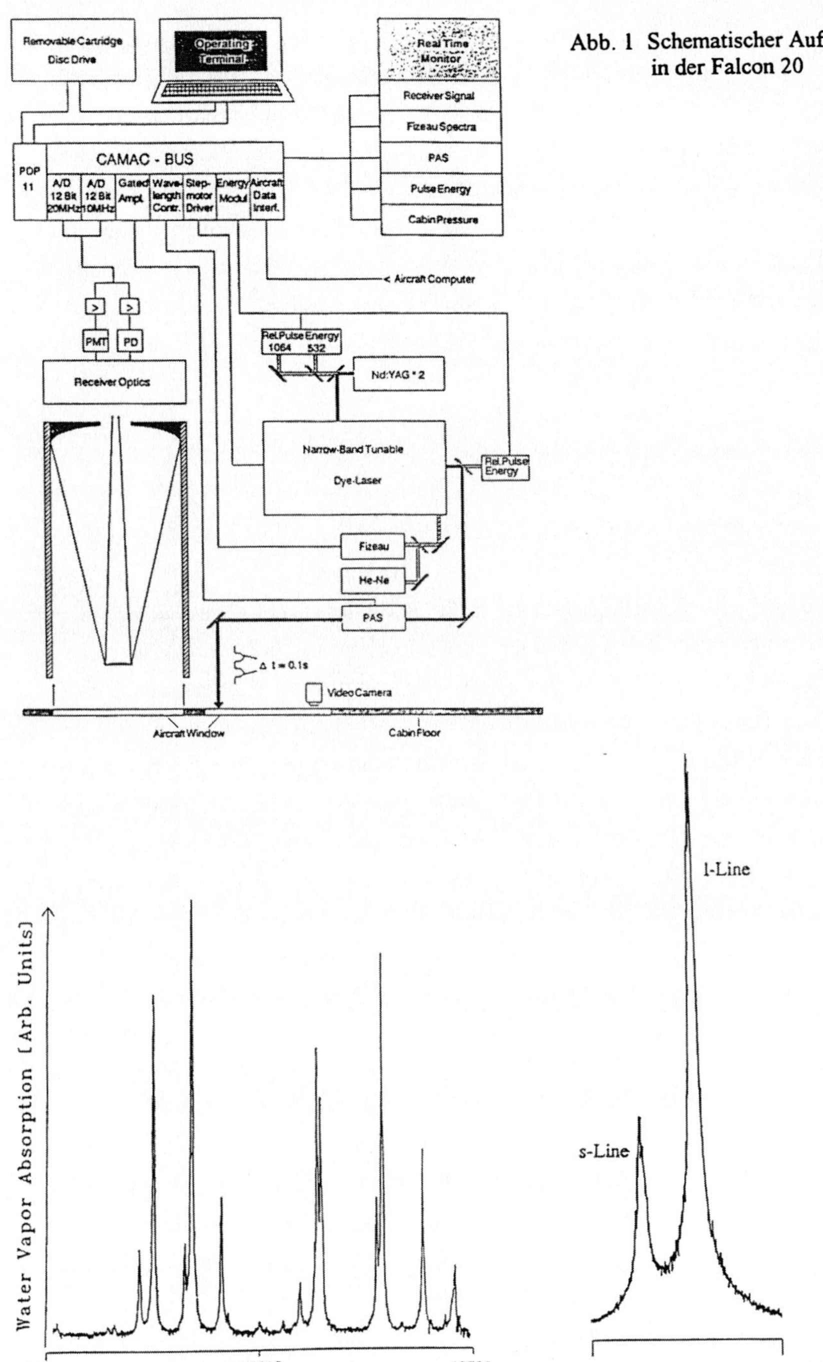

Abb. 1 Schematischer Aufbau des H_2O-DIALs in der Falcon 20

Abb. 2 Photoakustisches Spektrum von Wasserdampf im nahen Infraroten um 724 nm

zurückgestreuten Anteile von Laserpulsen zweier eng benachbarter Wellenlängen in unmittelbarer Nähe einer Absorptionslinie des zu messenden Gases. Bei der einen Wellenlänge wird viel Laserlicht absorbiert und schwächt damit die Laserstrahlung zusätzlich zu den Verlusten an den Luftmolekülen und Aerosolpartikeln. Die zweite "Referenzwellenlänge" wird so gewählt, daß eine deutlich geringere Absorption stattfindet. Aus den empfangenen Strahlungsströmen beider Wellenlängen läßt sich die Wasserdampfverteilung in Ausbreitungsrichtung des Lasers ermitteln. Das DIAL-Verfahren für Wasserdampfmessungen im nahen Infraroten ist in der Literatur ausführlich beschrieben (Schotland et al., 1974 u. C. Cahen et al., 1982).

Das DIAL-System

Das DLR Institut für Physik der Atmosphäre in Oberpfaffenhofen hat langjährige Erfahrungen beim Aufbau und der Anwendung von Lidargeräten im Flugzeug. Seit 1989 ist zur Wasserdampfprofilierung ein flugzeuggestütztes H_2O-DIAL im Einsatz. Es liefert die Wasserdampfkonzentration in vertikalen Schnitten der Atmosphäre mit hoher räumlicher und zeitlicher Auflösung. Abb. 1 zeigt eine schematische Anordnung des Gesamtaufbaus des DIALs eingebaut in der Falcon 20. Als abstimmbare gepulste Lichtquelle wird ein Farbstofflaser verwendet. Dieser wird von einem frequenzverdoppelten Nd:YAG-Laser bei einer Wellenlänge von 532 nm gepumpt. Der Farbstofflaser besteht aus einem Oszillator und zwei Verstärkerstufen. Im Oszillator wird die spektrale Intensitätsverteilung der Farbstofflaserpulse mit Hilfe eines Gitters und Etalons aktiv eingeengt. Man erreicht damit sehr kleine Laserbandbreiten ($\Delta\lambda$ < 1 pm). Diese schmalbandigen Pulse durchlaufen anschließend zwei Verstärkerstufen, womit sehr hohe Pulsleistungen erzielt werden (ca. 1 MW bei 720 nm). Die Pulsfolgefrequenz beträgt 10 Hz. Die Wellenlänge des Farbstofflasers läßt sich durch synchrones Verkippen der frequenzselektiven Elementen (Gitter und Etalon) im Oszillator in Schritten kleiner als 0,1 pm durchstimmen. Damit ist eine präzise Abstimmung der Sendefrequenz auf die Linienmitte einer Wasserdampflinie möglich. Das Verstimmen erfolgt mit rechnergesteuerten Schrittmotoren. Die Referenzwellenlänge erhält man durch sehr schnelles Verkippen des Gitters, bis der Oszillator auf dem nächsten Transmissionsmaximum des Etalons anschwingt. Während der Messung wird das Gitter mit einer Frequenz von ca. 4,5 Hz zwischen den beiden Wellenlängen hin und her verkippt. Die Wellenlängendifferenz zwischen der Absorptionsline und der Referenzlinie beträgt dann 39 pm. Die Wellenlängenkalibrierung und Überwachung erfolgt mit einer wasserdampfgefüllten, photoakustischen Zelle (Abb. 2) in Kombination mit einem Fizeau-Spektrometer. Damit kann die Sendewellenlänge während des Fluges auf ca. 0,3 pm genau kontrolliert werden.

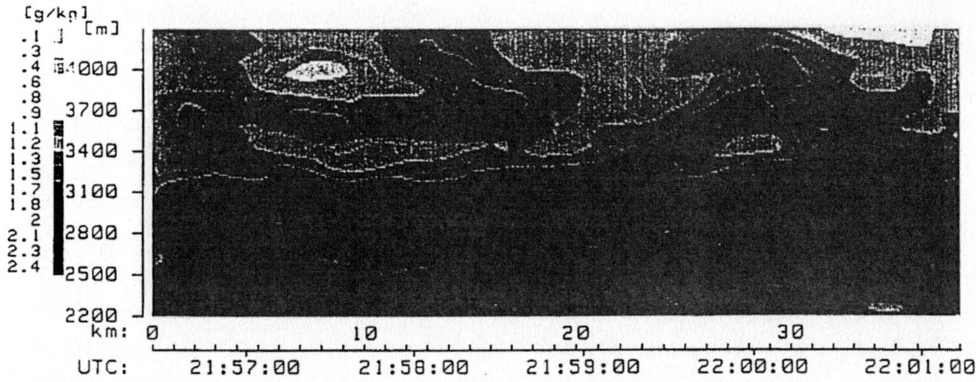

Abb. 3 Zweidimensionale Wasserdampfverteilung in der mittleren Troposphäre. Dargestellt ist das Massenmischungsverhältnis in g/kg in Abhängigkeit der Flugstrecke bzw. Flugzeit.

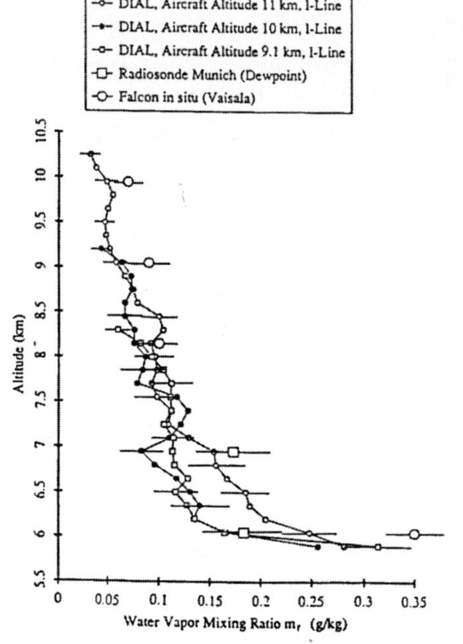

Abb. 4 Mittlere vertikale Wasserdampfprofile in Abhängigkeit der Flughöhen 11, 10, 9,1 km über demselben Meßgebiet. Gemittelt wurde über eine horizontale Flugstrecke von ca. 40 km. Für die DIAL-Messungen wurde die starke l-Linie aus Abb. 2 ausgewählt.

138

Der Empfangsteil besteht aus einem Spiegelteleskop (Cassegrain-Typ) mit einstellbarem Gesichtsfeld zur Unterdrückung der Hintergrundstrahlung. Damit können ohne Probleme Nachtmessungen durchgeführt werden. Bei Tage muß man jedoch ein schmalbandiges, temperaturstabilisiertes Interferenzfilter benutzen. Die mit dem Teleskop gesammelten rückgestreuten Photonen werden mit einem empfindlichen Photomultiplier nachgewiesen. Das Ausgangssignal des Photomultipliers wird anschließend verstärkt und mit einem schnellen Analog-Digital-Wandler digitalisiert und gespeichert. Die Steuerung des Gesamtsystems und die Überwachung der Datenaufzeichnung übernimmt ein schneller Kleinrechner. Eine ausführliche Beschreibung dieses Systems und eine quantitative Datenanalyse von Messungen in der unteren, mittleren und oberen Troposphäre ist bei Ehret et al. (1993) zu finden.

Ergebnisse

Abb. 3 zeigt eine zweidimensionale Wasserdampfverteilung von einem Testflug in Oberbayern zwischen Straubing und Allershausen im Juli 1990. Die gesamte Flugstrecke betrug etwa 60 Km. Man erkennt deutlich eine wellenförmige Struktur mit einer Wellenlänge von ca 14 km in der Feuchteverteilung zwischen 3100 und 4000 m Höhe. Die Amplitude beträgt etwa 20% des Mittelwertes.   Für die Berechnung dieser Feuchteverteilung wurden 100 Schußpaare horizontal (etwa 4 km horizontale Auflösung) und 20 Pixel vertikal (300 m vertikale Auflösung) gleitend gemittelt. Der statistische Fehler dieser Mittelwerte ist kleiner als 10%. Wir bemerken, daß während dieses Testfluges zum ersten Mal zuverlässige Wasserdampf-DIAL Messungen vom Flugzeug aus in der oberen Troposphäre durchgeführt werden konnten. Ein Vergleich mit den Ergebnissen von "In situ"- Sensoren in Abb. 4 zeigt recht gute Übereinstimmung, wenngleich eine mittlere systematische Abweichung von ca 20% zwischen den DIAL-Werten und den Falcon "In situ"-Daten zu beobachten ist.

Literatur

R. M. Schotland: Errors in the Lidar Measurement of Atmospheric Gases by Differential-Absorption. J. Appl. Met. 13, 71-77 (1974)

C. Cahen, G. Megie, and P. Flamant: Lidar Monitoring of Water Vapor Cycle in the Troposphere. J. Appl. Met. 21, 1506-1515 (1982)

G. Ehret, Ch. Kiemle, W. Renger and G. Simmet: Airborne Remote Sensing of Tropospheric Water Vapor Using a Near Infrared DIAL System: Appl. Optics, in press, (1993)

Lidar-Sondierung der arktischen Stratosphäre zur Bestimmung von Ozonprofilen und der Verteilung von Polar Stratosphärischen Wolken

M. Wirth, G.Ehret, P. Mörl, W. Renger
DLR, Institut für Physik der Atmosphäre
Münchener Str. 20, 82234 Weßling

1. Einleitung

Am Institut für Physik der Atmosphäre der DLR in Oberpfaffenhofen wurde in den letzten Jahren ein kombiniertes Aerosol- und Ozon-LIDAR, genannt OLEX für Ozon Lidar EXperiment, entwickelt. Es handelt sich dabei um eine flugzeuggetragene Ausführung, die es erlaubt Rückstreuprofile bei drei verschiedenen Wellenlängen (532 nm, 354 nm und 308 nm) aufzuzeichnen. Weiterhin kann bei einer Wellenlänge von 532 nm das Depolarisationsverhalten der Aerosolteilchen bestimmt werden. Im Rahmen des vom Bundesministeriums für Forschung und Technologie geförderten Ozonforschungs-Programms (OFP) war das OLEX-System während der letzten beiden Winter über 400 Flugstunden an Bord einer Transall der Bundeswehr im Einsatz.

Ein Ziel des OFP ist die Untersuchung der Ozonchemie der polaren Stratosphäre. Das OLEX ist eines von fünf Experimenten, die in der Transall installiert wurden, um einen möglichst großen Satz von sich ergänzenden Daten zu erfassen. Ein Teilziel ist die Untersuchung des Beitrags von Polaren Stratosphärischen Wolken (PSC's) zum Ozonabbau (siehe z. B. Crutzen und Arnold). Das LIDAR ermöglicht die Detektion dieser zum Teil nicht sichtbaren Wolken, die je nach chemischer Zusammensetzung und Form der Partikel in verschiedene Klassen eingeteilt werden. Aus der Kenntnis des Rückstreuverhaltens bei verschiedenen Wellenlängen, sowie der Depolarisation des zurückgestreuten Laserlichts lassen sich Rückschlüsse auf den Typ der PSC ziehen. Weiter erlaubt das OLEX-System die großräumige Vermessung der horizontalen und vertikalen Verteilung des stratosphärischen Aerosols. Mit Hilfe eines Aerosolmodells lassen sich aus diesen Daten optische Dicken, wie sie für Strahlungsmodelle und zur Korrektur anderer Fernerkundungsverfahren gebraucht werden sowie die Masse und Oberfläche der Aerosolpartikel, welches wichtige Eingangsparameter zu Simulation der heterogenen Chemie an diesen Teilchen sind, bestimmen. Die gleichzeitige Bestimmung von Ozon-Vertikalprofilen nach dem DIAL-Prinzip mit einer Auflösung ≤ 1 km erlaubt es, eine Korrelation von Aerosol und Ozon für verschiedene Höhenschichten herzustellen und damit, in Verbindung mit den Ergebnissen der anderen Experimente an Bord der Transall, den Einfluß der (Chlor-)Chemie auf die Ozonschicht von reinen Transportphänomenen zu trennen. Ein weiterer Vorteil des OLEX-Lidars gegenüber LIDAR-Bodenstationen und Ballonsonden ist die, durch den Flugzeugbetrieb ermöglichte, hohe Flexibilität des Einsatzes. Zum Beispiel erfolgte im letzten Winter ein kurzzeitiges Abdriften des Polarwirbels bis in mittlere Breiten, wobei es mit der Transall möglich, war diese Luftmassen von Spitzbergen bis nach Griechenland zu verfolgen und die Ozonchemie unter sich plötzlich ändernden Beleuchtungsverhältnissen zu studieren.

2. Experimenteller Aufbau und Datenauswertung

Der Aerosolteil des OLEX-Systems verwendet das frequenzverdoppelte und verdreifachte Licht eines Nd:YAG Lasers. Mit diesem System ist es möglich, gleichzeitig ca. 15 ns kurze Lichtpulse mit Energien von 100 mJ bei 532 nm und 60 mJ bei 354 nm zu erzeugen. Die Wiederholrate liegt bei 5 Hz. Zur Messung der Ozonkonzentration nach dem DIAL-Prinzip wird noch Licht einer Wellenlänge benötigt, welches von den Ozonmolekülen stark absorbiert wird. Dazu findet beim OLEX ein XeCl-Eximerlaser Anwendung, der Lichtpulse von 150 mJ Energie bei einer Wellenlänge von 308 nm erzeugt.

Das Empfangssystem besteht aus einem Cassegrain-Teleskop mit 35 cm Hauptspiegeldurchmesser und 5 m Brennweite. Nach dem Aufteilen des Lichts in die verschiedenen Empfangskanäle und Herausfiltern des Hintergrundlichts durch dielektrische Strahlteiler bzw. Filter erfolgt der Nachweis über Photovervielfacherröhren. Die Analogsignale werden mit Hilfe eines Vierkanaloszilloskops digitalisiert und zur späteren Analyse auf einer magneto-optischen Platte gespeichert.

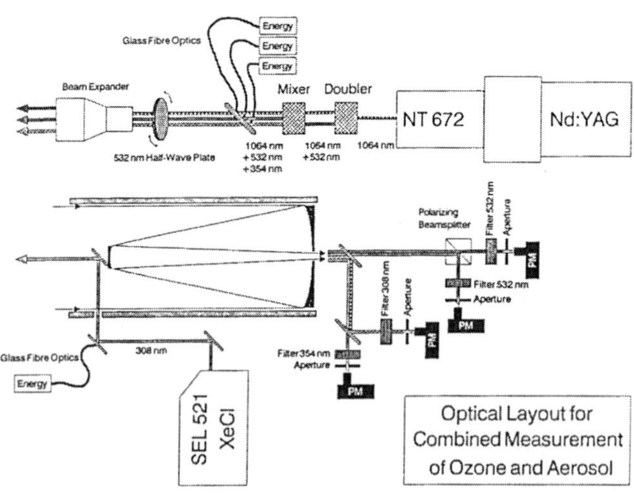

Bild 1: Das Bild zeigt schematisch den Aufbau des optischen Teils des OLEX-Systems. Im oberen Teil ist der Nd:YAG-Laser zu sehen. Über nichtlineare Kristalle wird Licht mit den Wellenlängen 532 nm und 354 nm erzeugt. Die $\lambda/2$-Platte dient zur Justage der Polarisationsebene des bei 532 nm emittierten Lichts. Die Strahlaufweitung reduziert die Divergenz des austretenden Lichtstrahls auf 0,5 m rad. Im unteren Teil ist der XeCl-Eximerlaser zu sehen, dessen Licht koaxial zum Teleskop ausgekoppelt wird. Der Nachweis arbeitet mit dielektrischen Strahlteilern und Filtern, sowie Photomultipliern als lichtempfindlichem Element.

Die vertikale Auflösung des Systems beträgt 12 m, die horizontale etwa 25 m für ein einzelnes Rückstreusignal bei einer typischen Fluggeschwindigkeit von 125 m/s. Das Signal/Rausch Verhältnis eines einzelnen Schusses genügt zur Bestimmung geometrischer Parameter, wie der Dicke und der Höhe von Aerosolschichten oder der Wellenlänge von Leewellen-PSC's. Für eine präzise Bestimmung des Rückstreukoeffizienten müssen ca. 1000 Einzelsignale aufaddiert werden, was einer horizontalen Auflösung von 25 km entspricht. Zur Bestimmung des Ozonprofils ist noch ein deutlich rauschfreieres Signal notwendig, so daß typischerweise über 10000 Signale gemittelt werden. Damit ergibt sich alle 250 km ein Ozonprofil mit einer Vertikalauflösung von 1 km. Der überstrichene Höhenbereich reicht von ca. 10 km bis 27 km geographischer Höhe.

Zur Berechnung von optischen Dicken, zur Extinktionskorrektur der Rückstreusignale und zur Berechnung des Aerosoleinflußes auf die Ozonprofile wurde ein auf Messungen von T. Deshler basierendes Aerosolmodell entwickelt und Mie-Rechnungen durchgeführt. Die sich ergebenden optischen Dicken und LIDAR-Verhältnisse stimmen recht gut mit SAGE II-Messungen (McCormik und Veiga) bzw. Raman-LIDAR Messungen (Ferrare et. al.) überein.

3. Ein Beispiel vom 25. März 93

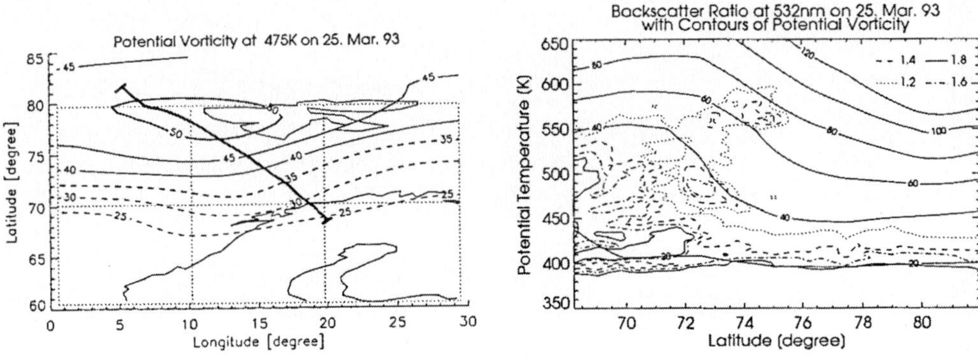

Bild 2: Die Darstellung links zeigt einen Ausschnitt eines Transallfluges vom 25. März 1993 von Kiruna nach Spitzbergen und zurück. Die dicke Linie kennzeichnet den Weg des Flugzeuges. Gleichzeitig sind Isolinien der Potentiellen Vortizität (in Einheiten von $10^{-6}\frac{K\,m^2}{kg\,s}$) auf einer Fläche konstanter potentieller Temperatur eingetragen. Daran läßt sich erkennen, daß bei diesem Flug der Rand des polaren Vortex gekreuzt wurde. Im Bild rechts sind Isolinien des vom OLEX-Lidar gemessenen Rückstreuverhältnisses bei einer Wellenlänge von 532 nm längs des Flugwegs aufgetragen (dargestellt über der geographischen Breite). Darüber gelegt sind wieder Isolinien der Potentiellen Vortizität. Gut zu erkennen ist die Aerosolmauer, die sich am Wirbelrand aufgebaut hat, während der Wirbel oberhalb von 420 K ($\approx$ 15 km) frei von Aerosol ist.

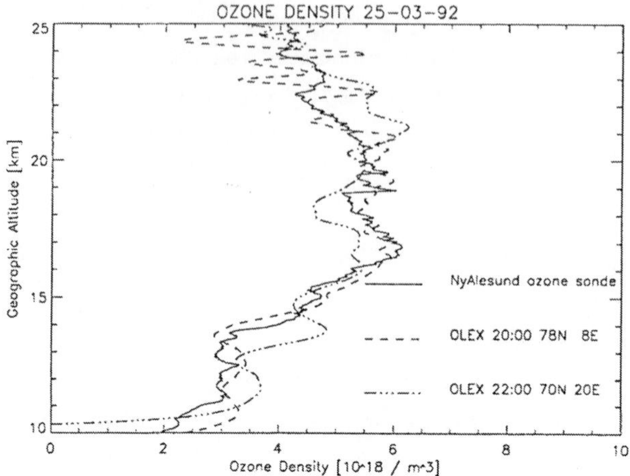

Bild 3: Bei dem in Bild 2 dargestellten Flug fanden auf dem Rückweg von Spitzbergen die Ozonmessungen statt. Das Bild oben zeigt zwei Ozonprofile, einmal im Wirbel (78°N) und ausserhalb des Wirbels (70°N). Weiterhin ist das Profil einer Ozonsonde des Alfred-Wegener Instituts eingezeichnet, die um 16^{00} Uhr in Ny Alesund auf Spitzbergen gestartet wurde. Man sieht eine gute Übereinstimmung der Ballon- mit der OLEX-Messung innerhalb des Wirbels, aber deutliche Abweichnungen bei dem am Rande des Wirbels aufgenommenen Profil. (Oberhalb von 20 km ist der Signal/Rausch-Abstand der 20^{00} Uhr OLEX-Messung schlecht, da noch keine völlige Dunkelheit herrschte.)

4. Ausblick

Als wesentliche Änderung soll bis zum nächsten Winter ein neuer Nd:YAG-Laser in das System integriert werden, der die fünffache mittlere Leistung bei 354 nm erbringt. Dadurch wird es möglich sein die Horizontalauflösung auf ca. 50 km herunterzudrücken oder die Genauigkeit des Systems im Bereich über 20 km zu verdoppeln.

Wir danken dem ECMWF für das Zurverfügungstellen der meteorologischen Analysen und dem AWI/Uni Bremen für die speziell zu Vergleichszwecken gestartete Ozonsonde.

Referenzen

Crutzen P. J. and Arnold F., Nitric acid cloud formation in the cold Antarctic Stratosphere: a major cause for the springtime ozone hole, *Nature, 324, 651-654, 1986*

Deshler T., In situ measurements of the size distribution of the Pinatubo aerosol over Kiruna on four days between 18 January and 13 Febuary 1992, *Geophys. Res. Lett.,special issue*

Ferrare R. A., Melfi S. H., Whiteman D. N. and Evans K. D., Raman LIDAR measurements of Pinatubo aerosols over southeastern Kansas during November-December 1991, *Geophys. Res. Lett., 19, 1599-1602, 1992*

McCormick M. P. and Veiga R. E., SAGE II measurements of early Pinatubo aerosols, *Geophys. Res. Lett., 19, 155-158,1992.*

Multiple Scattering and Depolarisation:
Classification of Ice and Water Clouds and PSCs

W. Krichbaumer
DLR, Institute of Optoelectronics
D-82230 Wessling
Postfach 1116

1   Introduction

For many years backscatter lidars have been used for remote sensing of the atmosphere. Although the determination of atmospheric parameters in thin aerosol by lidar is relatively easy, it is well known that the situation in dense atmosphere is more complicated. The reason for that is that in clouds or fog one always has to be aware of multiple scattering, whereas the usual evaluation procedures for the signals take only single scattering into account. Multiple scattering distorts the signal in the receiver; the inversion results achieved by the Klett method and other widely used algorithms are no longer correct. Hence multiple scattering is looked upon as a *perturbation effect*.

On the other hand, obviously a lot of information on the cloud or fog under consideration must be contained in the multiply scattered radiation: As the path of multiply scattered photons may contain any scattering angle between $0°$ and $180°$, the entire phase function is involved in the scattering mechanism, whereas backscattered light carries only information on the phase function at about $180°$. It follows that multiple scattering can be an additional *source of information* if methods are available to extract this information.

To achieve this it is necessary

a)   to have a lidar equipment designed *especially* for the measurement of the three main effects of multiple scattering:

- beam spreading
- pulse lengthening
- depolarisation,

b)   to make theoretical studies to improve the understanding of the mechanisms of scattering in clouds and fogs, and,

c)   based on these studies, to develop methods for evaluation of the measured signals under conditions of multiple scattering.

2   Instrumentation and Measurements

In recent years we have designed and built a new type of cloud lidar - the four-channel DLR-Microlidar with two polarisation components and two fields of view for each polarisation component - to meet these requirements. Technical details of the Microlidar concept have been reported before (e.g. Dahn et al., 1989; Werner et al., 1990; Werner et al., 1992). Figure 1 shows a schematic set-up of the transmitter/receiver unit.

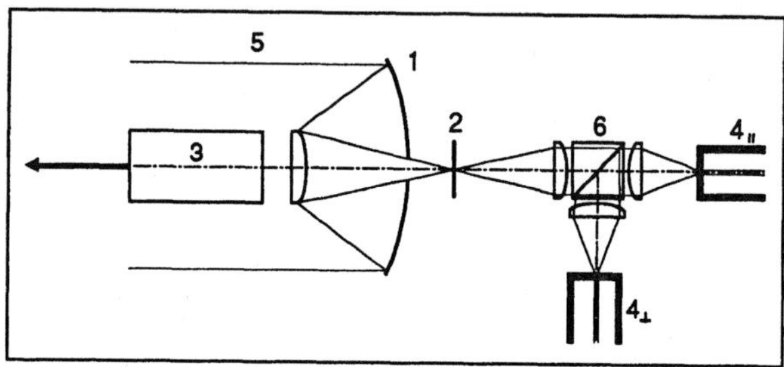

Figure 1   Schematic set-up of the Microlidar (1 parabolic mirror, 2 field stop, 3 diode - pumped Nd:YAG-laser, 4 fibres adjusted to different fields of view, 5 receiver field of view, 6 polarising beam splitter)

The transmitter is a modification of the newly-developed linearly polarised Nd:YAG laser which was introduced at "Laser 91" (cf. Mehnert et al., 1992). The system is capable of emitting pulses with 2mJ of energy at repetition rates up to 100pps requiring only passive, conductive cooling. A pulse duration of 12ns has been achieved, which is short enough for range-resolved lidar measurements. The output pulse energy can be well controlled from laser threshold up to maximum power. This enables us to make measurements both within dense clouds (minimal output power to avoid oversaturation of the detectors) and from some hundred meters above cloud top (maximal output power to achieve maximal signal amplitude).

The received signal is split up into the parallel and orthogonal polarisation components by a 50% beam-splitter in the receiver. The field of view (FOV) for each of the two polarisation components is separated into two regions (central part @ 3mrad half angle and ring-shaped exterior field of view @ 9mrad half angle) by custom-made fibre optics, thus ending up at four separate signals to be detected simultaneously. The exterior field of view for each polarisation component can (in the far field of the receiver) only contain contributions of multiple scattering, not of direct backscatter. In the near field this is no longer true, cf. Geinitz et al., 1993. A typical far field signal from a downward-looking airborne measurement in a dense mixed-phase cloud is shown in Fig. 2.

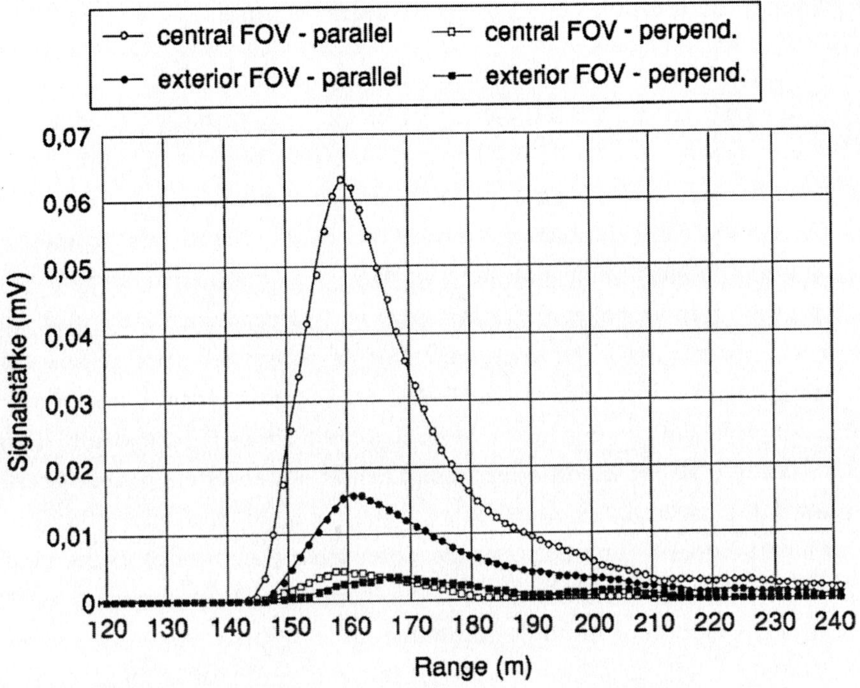

Figure 2 A four-channel Microlidar signal showing both multiple scattering and depolarisation

That this cloud does indeed contain ice particles can be seen immediately from the fact that the maximum of the perpendicular component in the central FOV (which contains all of the directly backscattered depolarised radiation) is at the same range as the maximum of the parallel component in the same FOV. Therefore considerable depolarisation by single scattering must have been occurred, and this is only possible with non-spherical particles. The other mechanism which can cause depolarisation of the lidar signal is multiple scattering, and this effect is a cumulative one, increasing with penetration of the laser beam into the cloud. This effect can be seen in the perpendicular component of the exterior FOV (notice the delayed maximum). The non-depolarised component of the signal is measured in the parallel channel of the exterior FOV. Also here a small delay of the maximum is present, caused by pulse lengthening on account of multiple scattering. Knollenberg measurements from the same cloud confirm this interpretation. Thus, a qualitative interpretation of the signals of the Microlidar is often immediately possible; however, for a quantitative analysis a complete theory of multiple scattering of laser pulses in clouds and fog is needed.

3   Theory of Multiple Scattering of Lidar Signals

In a cooperation with the Institute of Mathematics of Munich University (LMU) we constructed a theoretical model of multiple scattering of laser radiation by considering the flux of light as the stochastic process of propagation of (non-quantum-mechanical) photons. We derived general and exact analytical equations for the contributions of the various orders of scattering to the entire signal in the different channels of the lidar. Details of the mathematical methods can be found e.g. in Oppel et al., 1989, Oppel and Krichbaumer, 1990, and Krichbaumer and Oppel, 1990. The implementation of polarisation within this model has been investigated (Krichbaumer, 1989). Based on this concept, we solved the problem of polarisation anisotropy (Krichbaumer, 1990). It turned out that, in fact, this is not a multiple scattering phenomenon but an effect of the sphericity of the scattering particles.

In 1991 a cooperation between the Computing Centre (CC) of the Russian Academy of Sciences / Novosibirsk and the Institute of Mathematics of LMU in the field of calculation of multiple scattering started. This cooperation as well as the collaboration with DLR was within the frame of a project sponsored by DFG (German Science Foundation). As a result, the different approaches of DLR / LMU and of CC could be unified to a great extent; a joint, very powerful computer code for the calculation of lidar signals of arbitrary orders of scattering was developed and successfully tested at LMU, cf. Oppel et al., 1993. With this model, a well-understood and sophisticated theoretical tool for the interpretation of multiple scattering phenomena is now at hand.

4   Interpretation of Measured Lidar Signals

The exact equations for the lidar signals for various orders of scattering based on the mathematical model above are rather complicated and often not very useful for practical evaluation of signals. However, in most situations, e.g. with respect to geometry of measurement or to meteorological conditions, we are able to replace them by much simpler formulae derived from the exact ones and which are a good approximation to these.

We do not want to go into a more detailed analysis here but only demonstrate briefly that these approximations are well in agreement with the experimental results provided they are applied in the correct situation.

Consider the situation of a cloud in some distance below the laser, as it is the case e.g. in the measurement example of Fig. 1. The cloud is assumed to consist of piecewise (more or less) homogeneous layers. Let us denote the single-scattering signal by μ_1 and the double-scattering signal by μ_2. It can be shown that in such a homogeneous layer

a) the logarithm of the r^2-corrected single-scattering contribution (i.e. $\log[r^2 \mu_1(r)]$) is a straight line with slope proportional to σ (extinction coefficient);

b) the logarithm of the r-corrected double-scattering contribution (i.e. $\log[r \mu_2(r)]$) is a straight line with slope proportional to σ; and that

c) the ratio $\mu_2(r)/\mu_1(r)$ of both signals forms a straight line with slope σ/β.

a) is the traditional slope-method. If we accept the hypothesis that the signal in the central field of view for each polarisation direction contains only single-scattering (i.e. is equal to μ_1) and the signal in the outer fields of view contains only double-scattering (i.e. μ_2) then we should find these three straight lines in the measured signals. And this is the case: Figs. 3 and 4 show the pertinent lines from a) to c) above, evaluated from the measured signals of Fig. 2.

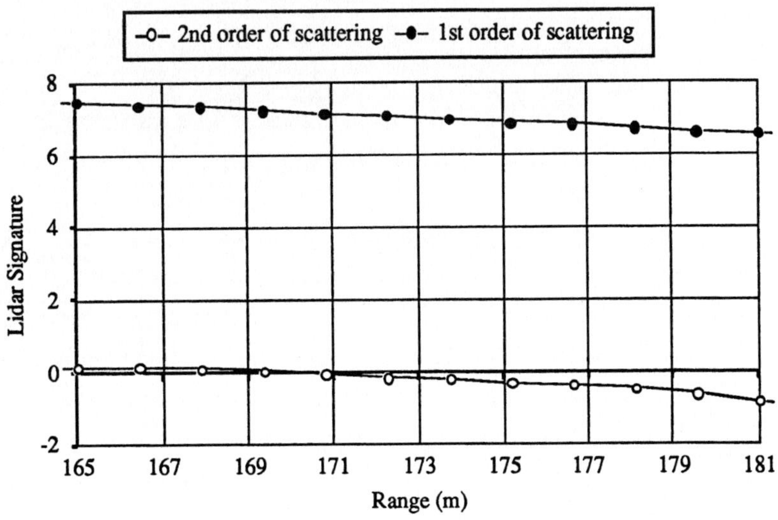

Fig.3: Slope Method for Double and Single Scattering

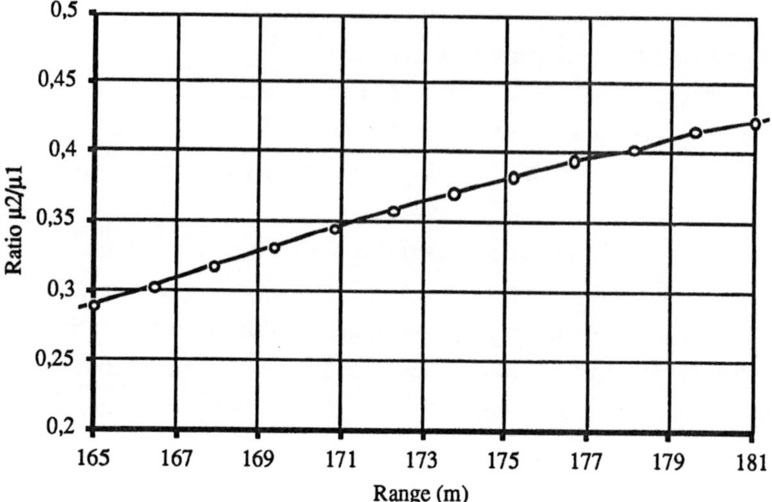

Fig.4: Ratio Double to Single Scattering

Naturally these hypotheses and assumptions are not valid in all measurement situations: there could be higher orders of scattering than the first in the central fields of view, or higher ones than the second in the exterior field of view, and so on. However, the considerations above should only serve as an example that theory and experiment are well suited to each other; of course, in more complex cases we would apply more sophisticated methods.

5   Future Developments

The Microlidar concept for lidar measurements in dense clouds and fogs, together with the theoretical tools, is well integrated into major national and international activities. Some examples:

- CEC-project "Improvement of Lidar Measurement Techniques for Discrimination of Polar Stratospheric Clouds and Volcanic Aerosols" together with RISØ (Denmark) and IROE-CNR (Italy). This project is dedicated to the improvement of particle measurement techniques with lidars (ground-based and airborne) to discriminate between various types of particles (aerosols, polar stratospheric clouds etc.) The special emphasis is on the study of optically dense media, i. e. tropospheric and stratospheric clouds.

- For measurements on the stratospheric aircraft Strato 2C, a multi-purpose lidar ADLER was proposed. The DIAL off-line channel will be modified as a four-channel backscatter lidar according to the prototype concept of the Microlidar.

According to theoretical predictions, the contribution of multiple scattering to the total lidar signal will increase with distance to the cloud. Therefore a four-channel backscatter lidar will be of special interest for spaceborne applications. For use on Russia's MIR-2 or ALMAZ, a spaceborne lidar based on the prototype concept of the Microlidar has been proposed.

Literature:

Dahn, H.-G. (1989): Experimental Measurement of Multiple Scattering and Depolarisation with the Microlidar, Proc. MUSCLE 3, 3rd International Workshop on Multiple Scattering Lidar Experiments, Oberpfaffenhofen (Germany), Oct. 24-26, 1989.

Geinitz, V., Richter, W., and Krichbaumer, W. (1993): Near-Field Effects in a Monostatic Multiple-Aperture Lidar. These proceedings.

Krichbaumer, W. (1989): Polarisation and the Stochastic Model. Proc. MUSCLE 3, 3rd International Workshop on Multiple Scattering Lidar Experiments, Oberpfaffenhofen (Germany), Oct. 24-26, 1989.

Krichbaumer, W. (1990): On the Origin of Polarisation Anisotropy, Proc. MUSCLE 4, 4th International Workshop on Multiple Scattering Lidar Experiments, Florence (Italy), Oct. 29-31, 1990.

Krichbaumer, W., and Oppel, U.G. (1990): Relation of multiply scattered return signals to the interpretation of airborne measurements with the DLR-Microlidar,Proc. MUSCLE 4, 4th International Workshop on Multiple Scattering Lidar Experiments, Florence (Italy), Oct. 29-31, 1990.

Mehnert, A., Peuser, P., and Schmitt, N. (1992): New Solid State Lasers for Applications in Lidar Systems. In.: Laser in der Umweltmeßtechnik (Eds. Ch. Werner, V. Klein, K. Weber). Springer-Verlag Berlin 1992, 201-205.

Oppel, U.G., Findling, A., Krichbaumer, W., Krieglmeier, S., and Noormohammadian, M. (1989): A Stochastic Model for the Calculation of Multiply Scattered Lidar Returns. DLR-Forschungsbericht DLR-FB 89-36.

Oppel, U.G., and Krichbaumer, W. (1990): Calculations of multiply scattered lidar return signals from slabs of dense aerosols, Proc. MUSCLE 4, 4th International Workshop on Multiple Scattering Lidar Experiments, Florence (Italy), Oct. 29-31, 1990.

Oppel, U.G., Starkov, A.V., and Noormohammadian, M. (1993): A Stochastic Model and a Variance Reduction Monte Carlo Method for the Calculation of Light Transport. To appear in Appl. Phys. B.

Werner, Ch., Hörmann, P., Dahn, H.-G., and Herrmann, H. (1990): Technical Problems with Respect to the Separation of Single and Multiple Scattering in a Monostatic Lidar, Proc. MUSCLE 4, 4th International Workshop on Multiple Scattering Lidar Experiments, Florence (Italy), Oct. 29-31, 1990.

Werner, Ch., Streicher, J., Herrmann, H., and Dahn, H. G. (1992): Multiple-Scattering Lidar Experiments. Opt Eng 31 (1992), 1731-1745.

Doppler Lidar for Atmospheric Remote Sensing and Aircraft Safety Operations

Friedrich Köpp

DLR-Institute of Optoelectronics, D-82230 Oberpfaffenhofen

1. Introduction

The measurement of the wind field is widely recognized as fundamental for the accurate description of atmospheric dynamics and hence for the understanding and prediction of weather development. It has taken many decades for direct wind sensors like cup-anemometers and balloon sondes to be established. Now, they are partly replaced by remote sensors based on acoustic and electromagnetic radiation. The most promising remote sensing method for wind and turbulence measurements over various scales is the Doppler Lidar technique. At the DLR-Institute of Optoelectronics, there exists a long-standing experience in the development and application of ground-based systems which are based on cw, CO_2 lasers. Pulsed Doppler lidars for airborne applications are presently under development.

2. The Doppler Lidar technique

The heterodyne lidar consists of: a continuous wave (cw) or pulsed frequency-controlled laser transmitter locked to a local oscillator laser, a transceiver telescope, a heterodyne detector where the local oscillator radiation is mixed with the Doppler-shifted backscatter signal, and a signal processing system. The signal originates from the radiation backscattered by small aerosol particles which move with the prevailing wind speed and direction through the sensing volume. In the case of cw lasers, the sensing volume is defined by the depth of focus of the telescope, otherwise by the length and position of the laser pulse. The Doppler frequency shift Δf_D directly determines the line-of-sight (LOS) component V_{LOS} of the wind vector:

$$\Delta f_D = 2 \cdot (V_{LOS}/c) \cdot f_0 \tag{1}$$

where c is the light velocity and f_0 the laser frequency. At the CO_2 laser wavelength

(λ = 10.6 μm) a velocity component of 1 m/s corresponds to a frequency shift of 189 kHz. A measurement at an azimuth angle θ and an elevation φ yields a V_{LOS} which is related to the wind vector components u, v and w:

$$V_{LOS} = u \sin\theta \cos\varphi + v \cos\theta \cos\varphi + w \sin\varphi \qquad (2)$$

From the V_{LOS} components measured during a full azimuth rotation at constant range and elevation the wind vector in one atmospheric layer can be determined by using the VAD (velocity azimuth display) method. For a homogeneous flow V_{LOS} versus θ follows a sine-wave shape. The data of the measurements in a turbulent atmosphere have to be fitted to a sine wave from which the vector components u, v and w can be derived. By repeating that evaluation procedure for each layer, the vertical profile of wind vectors is built.

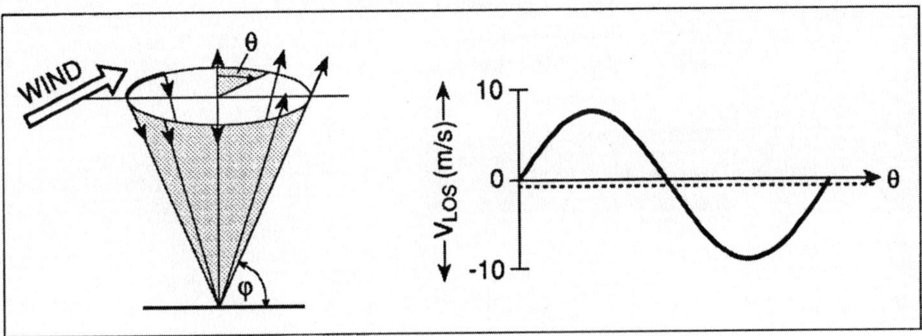

Figure 1: VAD method for wind vector determination [1]

3. The ground-based Laser Doppler Anemometer

The ground-based Laser Doppler Anemometer (LDA) has been developed for wind and turbulence investigations in the atmospheric boundary layer [2]. It is a continuous wave system based on a 4 Watt CO_2 laser (eye-safe wavelength of 10.6 μm) and a transceiver telescope of 30 cm diameter. By changing the focal length of the telescope, the system range can be varied between 40 and 1000 m. The flexible scanning device enables one to point the measuring beam in all directions, e.g. for performing the VAD scan for wind vector determination. The control of the scan procedures, the data acquisition and the on-line data evaluation is performed by a Compaq 386. Some system properties are listed in Table 1.

Table 1:   Laser Doppler Anemometer (LDA) system parameters

Laser: CO_2, continuous wave

output power	4	W
short-term stability	3	kHz

Transceiver telescope:

aperture	30	cm
focal length	90	cm

Detector: HgCdTe, cooled to 77 K

sensitive area diameter	0.5	mm
detectivity	$5.8 \cdot 10^{10}$	cm $(Hz)^{1/2}$ / W

Range resolution:

at 100 m range	6	m
at 500 m range	150	m

Spectrum analyzer: surface acoustic wave

frequency resolution	20	kHz
corresponding velocity resolution	0.1	m / s

Scanning device:

angle resolution	0.1	degree

Acquisition of frequency spectra:

maximum storage rate	160	Hz

One fundamental task for a wind sensor is the determination of vertical wind profiles. The potential of the VAD method for that task has been demonstrated during
various field experiments where balloon-borne wind-sondes were launched simultaneously with the LDA measurements [3]. The observed root-mean-square (RMS)
deviations did not exceed 1 m/s, despite the differences of both sensing techniques: e.g. the lidar velocities are determined by the Doppler shift whereas the
sonde velocities are derived from changes in the balloon position, the lidar sensing
volume is centered above the scanner whereas the balloon drifts with the wind so
that the data are acquired from different atmospheric regions that may have very
different wind values. The big advantage of the method is the ability to obtain volume-averaged, three-dimensional wind-vector profiles in a short measurement
time and with a high repetition rate. The results were verified by experimental investigations in USA using the pulsed Doppler Lidar of NOAA [4].
In the ground-based version the cw Doppler Lidar is best suited for the investigation of small-scale phenomena like the flow field in the vicinity of emitting sources,
the exchange of pollutants, the wind and turbulence profiles in the atmospheric

boundary layer, and-so-on. Moreover, it has proved to be an important tool for experimental investigation of aircraft wake vortices.

Wake vortices are a severe air-traffic problem. Especially during landing and take-off, the strong vortices of heavy aircraft present a potential hazard to other aircraft following closely behind [5]. Vortex imposed separation standards have since been established. But they have proved to be a strong restriction on the capacity demands at major airports. The German Wake Vortex Program is centered on the highly frequented Frankfurt/Main Airport. At that airport additional capacity limitations are necessary due to the separation of 518 m between two parallel runways, a separation which is too small for operating both runways independently with respect to the effects of wake vortices.

During several extended field experiments in the years 1983-85 and 1989-90 the LDA has been operated at Frankfurt/Main measuring the vortices of more than 1400 landing aircraft of both a heavy and large variety [6]. The LDA container was positioned between the landing corridors of runways 25R and 25L, about 850 m in front of the runway thresholds. As sketched in Figure 2, a section of the vertical measurement plane across one of the approach corridors is covered by a fast elevation scan at fixed range setting. The right side of the figure shows the measured velocity profile, here the profile of a B-747 port vortex. This profile is the superposition of the vortex rotational field and the lateral vortex transport including the cross-wind component. Even though the LDA is operated in homodyne mode, it is possible to distinguish positive and negative velocity components by using the known cross-wind direction for reference. In this case, the profile branch above 35 m altitude has to be tilted to negative velocities. After the vortex has passed that sensing region, the next one is chosen by changing the range setting. As soon as the vortex has reached the LDA position, the measurement half-plane is turned in azimuth by 180° and the vortex tracking is continued towards the parallel runway. By means of that "gate method" the vortex transport between the parallel runway system can be investigated.

In Figure 3 the results of a complete measurement sequence are compiled which is the landing of a B-747 on runway 25L. In the upper part, the maximum tangential velocities are shown; the vortices ages (in seconds) are labelled by the numbers. The lower part shows the positions of the downwind vortex during the transport extending over the LDA container towards the parallel runway. Driven by the cross wind, the vortex descends on a slant slope to ground proximity. There, the lateral

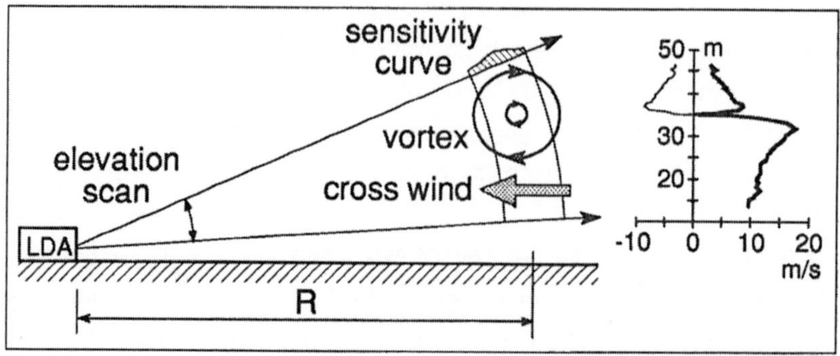

Figure 2: Strategy of the wake vortex measurements using the LDA system

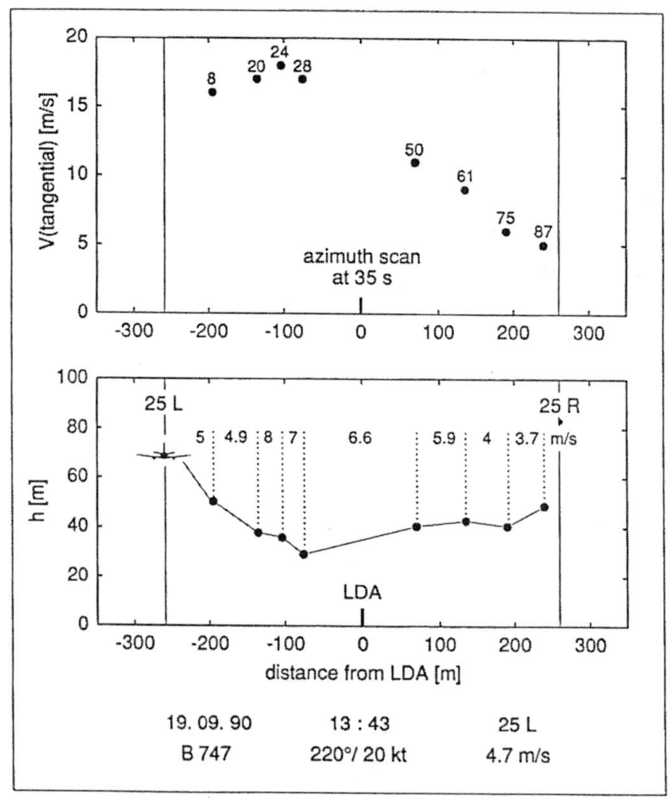

Figure 3: Results of one complete measurement sequence, showing vortex
strength (upper part) and vortex location (lower part).

motion is intensified by the self-induced velocity. In this example, the horizontal shift velocity is at a maximum value of 8 m/s directly after descent and approaches the crosswind value of close to 4 m/s after 61 s. Under crosswind conditions of that amount, the vortices actually reach the safety area of the parallel runway with high probability, in this case after 87 s still exhibiting critical strength.

4. Airborne Doppler Lidar systems

The German-French project WIND

The transition from smallscale to mesoscale investigations can be realized by pulsed airborne systems.

A pulsed CO_2 Doppler Lidar is under development in a German-French cooperation involving DLR and CNES/CNRS. This project is called WIND, which is the acronym of Wind INfrared Doppler lidar. It is characterized by two major goals: the investigation of mesoscale wind fields over land and over sea and the gathering of experience for the development of a spaceborne system.

The results of the extensive Phase-A study are comprised in the WIND-Phase-A-Report [7]. There, the state of knowledge concerning the possible applications of an airborne Doppler Lidar and the technological background for the development of such a system is described. During Phase B the design criteria of the WIND system were elaborated and the critical components were investigated [8]. Instrument construction and aircraft installation shall be completed by the end of 1995 with first test flights and scientific campaigns being scheduled for 1996.

The main components of the WIND system are sketched in Figure 4. The platform is the DLR research aircraft Falcon 20 allowing 3 hours of operation at an altitude of 10 km. The TE and LO laser package, the opical bench including several detectors for heterodyne detection and monitoring, the transceiver telescope of 20 cm diameter and the Ge wedge for performing nadir-looking conical scans with 30° half-cone angle are integrated in the first mounting frame which is installed above the Ge window on the aircraft floor. A second mounting frame contains the subsystems for data processing, display, and storage, and the electronics control. The latter comprises the instrument house-keeping and the aircraft parameters necessary for data correction. The specifications of WIND are summarized in Table 2.

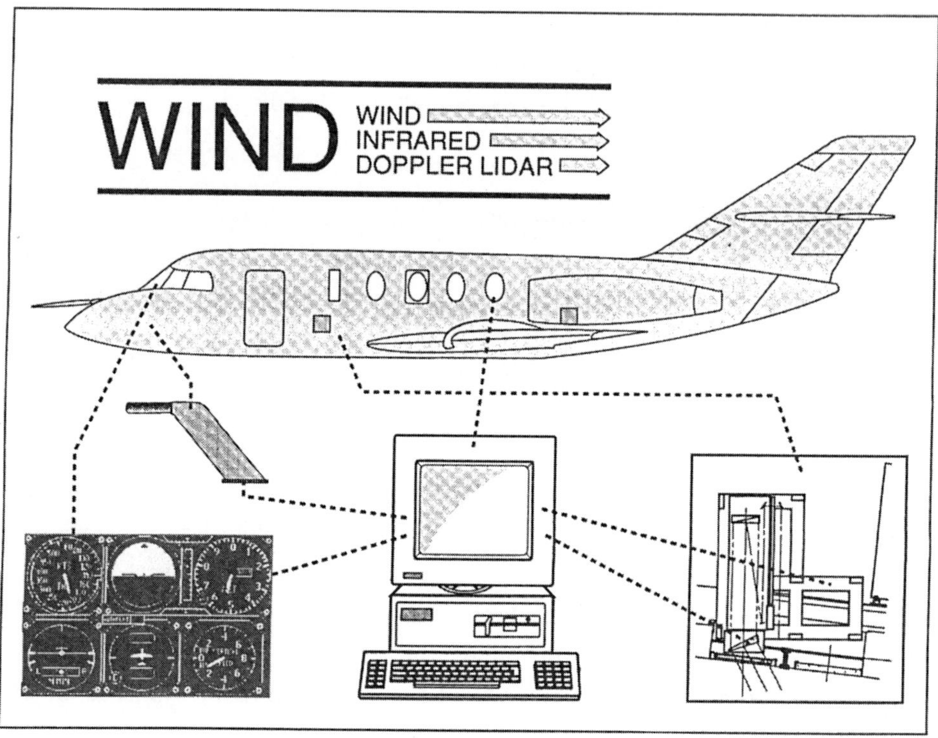

Figure 4: Subsystems of the airborne Doppler Lidar WIND.

Most of the possible applications of WIND are directed specifically to mesoscale processes relevant to meteorology and climate, in connection with the variability of the continental surface, the flow perturbations by orographical effects, the ocean/atmosphere interaction, the dynamics of convective and stratiform cloud systems, and the moisture convergence. The mesoscale studies will also be beneficial for the larger scales by a better parametrization of the sub-grid variance of synoptic models.

Table 2: Parameters of the WIND System

CO_2 TE Laser:	wavelength	10.6	μm
	pulse energy	300	mJ
	pulse length	3	μs
	pulse repetition rate	10	Hz
	chirp	≤ 1	MHz
	shot-to-shot stability	≤ 1	MHz
	life time	≥ 10^7	shots
Transceiver telescope:	Dall-Kirkham, off axis		
	aperture	200	mm
Scanner (Ge-wedge):	scan angle (half cone)	30	°
	rotating rate	0.05	Hz
Signal processing:	concept: early digitizing		
	frequency estimator: PP, PPP		
Platform:	Falcon 20 aircraft		
	flight altitude	10	km
	aircraft speed at 10 km	170	m/s
Measurement geometry:	height resolution	250	m
	grid size	10 x 10	km^2
Measurement accuracy:	single LOS components	1 - 3	m/s
	total wind vector	0.5	m/s

The CEC Project FLAME

The DLR experience with wake-vortex measurements using the ground-based cw Laser Doppler Anemometer (LDA) and with the realization of the airborne pulsed Doppler Lidar WIND have been combined and included in the CEC Program FLAME. The program consortium consists of two industrial companies, two research establisments, and two university institutes. The partners and their main responsibilities are listed in Figure 5.

First simulations have been carried out and have led to the first conclusions:

- The development of a routine sensor for installation in commercial aircraft necessitates a small and robust design. Since the sensor size is depends on the transmitter wavelength, the transition from CO_2 (10 μm) to shorter wavelengths is envisaged. The problem is to find a frequency-stable solid state laser in the eyesave region (around 2 μm) of similar reliability to CO_2 lasers.

PARTNERS	CONTRIBUTIONS
Sextant Avionique Valence, France	project coordination, system analysis
GEC Avionics Rochester, UK	laser and optics technology, demonstrator realization
DLR Optoelectronics Wessling, Germany	vortex and lidar modelling, system simulations
University College Galway, Ireland	signal processing, pattern recognition
Universität Hamburg Hamburg, Germany	selection of laser crystals supply with laser crystals
INESC Institute Porto, Portugal	simulation of laser design, simulation of optics design

Figure 5: The consortium of the CEC Program FLAME

- The role of range resolution is controversial. Long range cells, generated by long laser pulses or by signal processing algorithms, result in high SNR values and hence in small velocity errors. On the other hand, the vortex extension is relatively small and, therefore, can be better detected using short range cells. A trade-off concerning the optimum range resolution is thus necessary.
- Due to noise and speckle effects, single measurements are not representative. Therefore, a fast scanning for generation of three-dimensional pictures with high repetition rate is recommended. That will enable the recognition of a vortex with the aid of an intelligent pattern recognition system.
- Since the airborne Doppler Lidar has to detect the vortices of an aircraft flying in front of the sensor, the axial flow component of the vortex will mainly provide the

characteristic signature. Until now, little information is available about the axial flow field. Therefore, theoretical and experimental investigations are necessary. At the end of the FLAME Program in July 1995, the possibility of detecting wake vortices by an airborne Doppler Lidar will be clarified and a ground-based demonstrator unit will be realized.

4. Summary

The Doppler Lidar is the most promising remote sensing method for wind and turbulence investigations. In ground-based cw version it is appropriate for precise boundary layer measurements within a range of 1000 m. That range can be increased to 10 - 20 km by using a pulsed system. Airborne pulsed Doppler Lidars like WIND and FLAME will extend the operational range to mesoscale meteorological investigations and to airsafety-relevant problems. For measurements in global scale spaceborne Doppler Lidars (ALADIN, LAWS) are considered by the space agencies.

[1]   Klein, V. and Ch. Werner, "Fernmessung von Luftverunreinigungen," Springer-Verlag, 1993.

[2]   Köpp, F., H. Herrmann, Ch. Werner, R.L. Schwiesow, and F. Bachstein, "Erstellung und Erprobung des Laser-Doppler-Anemomters," DFVLR-FB 83-11, 1983.

[3]   Köpp, F., R.L. Schwiesow, and Ch. Werner, "Remote measurement of boundary-layer wind profiles using a cw Doppler lidar," J. Climate and Appl. Meteor., 23, 148, 1984.

[4]   Hall, F.F., R.M. Huffaker, R.M. Hardesty, M.E. Jackson, T.R. Lawrence, M.J. Post, R.A. Richter, and B.F. Weber, "Wind Measurement Accuracy of the NOAA pulsed infrared Doppler Lidar," NOAA Tech. Memo, ERL WPL-37, 1987.

[5]   Hallock, J.N. and W.R. Eberle (Editors), "Aircraft wake vortices: A state-of-the-art review of the United States R&D Program," FAA-RD-77-23, February 1977.

[6]   Köpp, F., "Experimental investigation of aircraft wake structure and propagation using the DLR Laser Doppler Anemometer," Proceedings of Aircraft Wake Vortices Conference, Washington, D.C., October 1991.

[7]   Köpp, F. and C. Loth (Editors), "WIND, Phase-A Report," DLR/CNRS/CNES, December 1989.

[8]   Köpp, F. and C. Loth (Editors), "WIND, Phase-B Final Report," DLR/CNRS/ CNES, May 1993.

Proposed Measurements of the Global Wind Field Using Laser Doppler Method

Christian Werner
Institute of Optoelectronics
German Aerospace Research Establishment (DLR)
P.O.B. 1116   D-82230 Wessling

1. Introduction

The last ten years have seen a significant improvement in our ability to forecast the weather and simulate the general circulation of the atmosphere. This success has served to expose and underline the need for a significant improvement in the global and regional observing system. In this connection special mention should be made of observations of wind fields in three dimensions as these are of special importance to both operational weather forecast and studies of the Earth's climate. The height-resolved wind vector measurements from space in global coverage is not possible at present, but a conceptional design exists. The availability of this observation would have a dramatic impact on the quality of weather forecast. Complementary measurements will be provided near the surface where wind is a driving force. GEWEX (Global Energy and Water Cycle Experiment) has been initiated by the World Meteorological Organization. The estimation of energy convergence and divergence asks for the precise measurement of wind with the accuracy and resolution shown in table 1.

Table 1 Requirements for a spaceborne lidar

	Stratosphere	**Troposphere**
horizontal resolution	100 km (50 km)	100 km (50 km)
vertical resolution	3 km	1 km between 2 - 15 km 0.5 km below 2 km
accuracy	2 - 3 m/s	1 - 2 m/s
frequency per day	4	4

The design of a sensor and the problems to realise the measurements are reported in this paper.

2. Doppler Lidar Technique

A heterodyne lidar consists of a pulsed, frequency-controlled laser transmitter (L1) locked via a locking loop (LL) to the local oscillator (LO), a transmit and receive telescope , a heterodyne detector (D), where the local oscillator (LO) radiation is mixed with the Doppler-shifted backscatter signal , and a signal processing system. The signal origines from the backscattering by small aerosol particles which move with the prevailing windspeed through the laser focus volume. The windshifted Doppler frequency directly determines the line-of-sight (LOS) component (V_{LOS}) of the wind vector (Figure 1).

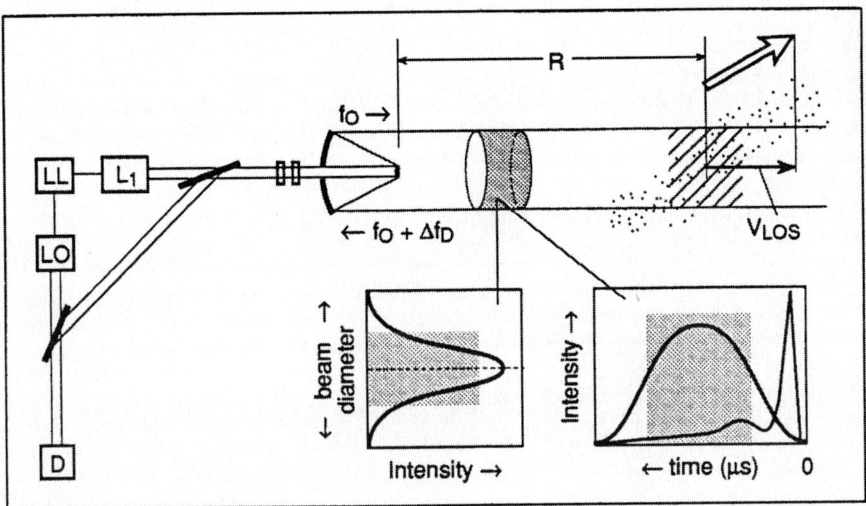

Fig. 1 Scheme of a Doppler lidar

The laser pulse with an optical carrier frequency f_0 has a length of a few microseconds and a Gaussian beam profile (Figure 1) (Post and Cupp 1990). **One measures the backscattered, Doppler shifted radiation.** Backscattering depends on the sizi and number density of the aerosol particles available in the volume of interest. **The backscattered Doppler frequency directly determines the line-of-sight (LOS) component (V_{LOS}) of the wind vector.** At CO_2 laser (wavelengths $\lambda = 10.6$ µm) a velocity component of 1 m/s corresponds to a frequency shift, Δf_D of 189 kHz. This is obtained from the equation

$$\Delta f_D = 2 \frac{V_{LOS}}{c} \cdot f_0 \qquad (1)$$

where c is the speed of light and V_{LOS} is the line-of-sight component.

By scanning the transceiver with a conical scan pattern (VAD) one reaches different azimuth and elevation angles to combine the single LOS-components to a wind vector (sine-wave-fitting). Conical scans performed from the airborne system WIND need a sector data handling and the spaceborne systems only LOS-components can be used.

By measuring at an azimuth angle θ and an elevation angle φ one gets a radial (line-of-sight) contribution V_{LOS} which depends on the wind vector components u, v, and w given by

$$V_{LOS} = u \sin\theta \cos\varphi + v \cos\theta \cos\varphi + w \sin\varphi \qquad (2)$$

It is accepted worldwide that ground-based Doppler lidars can measure wind profiles in the atmosphere. Various field experiments have been carried out where balloon wind-sondes were launched simultaneously with wind profile measurements made by Doppler lidar (Hall et al., 1984; Köpp et al., 1984). Radiosonde winds have been averaged over height ranges comparable to the lidar range resolution. The route-mean-square (RMS) differences are below 1 m/s. Lidar velocities are determined from a Doppler shift whereas sonde velocities are determined from changes in the balloon range. The lidar sensing volume is centered over the scanner on the average, but the balloon drifts with the wind so that the two techniques measure different regions of the atmosphere that may have different wind values.
Figure 2 shows the measurement concept of a VAD lidar.

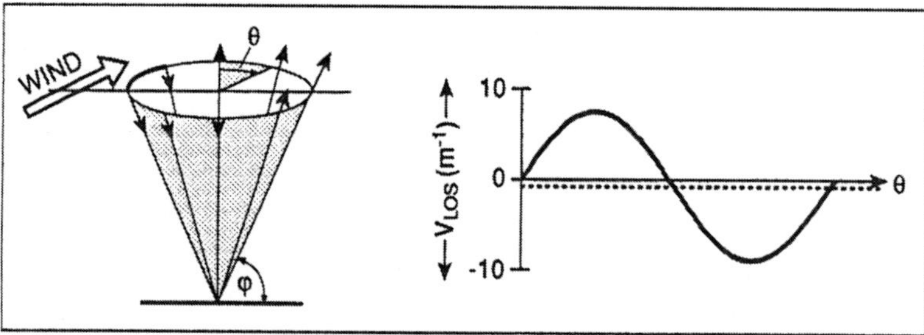

Fig. 2 Ground-based Doppler lidar operation

The conical scan technique is used. It is assumed that the wind is homogeneous over the scanning volume at each level, which is given by the lidar pulse length (range resolution). From equation 2 one obtains a set of LOS components with equal elevation angle φ and different azimuth angle θ. The task is to fit the measured radial velocity V_{LOS} versus azimuth data, using a standard least-square procedure to determine the vector components u, v and w. The right side of Figure 3 shows V_{LOS} versus azimuth angle. The deviation of the average from the zero line is directly proportional to the vertical wind component. (A negative (up) component is shown).

The great advantage of ground-based Doppler lidar when using the conical scan is the ability to obtain a volume-averaged, 3-dimensional vector wind profile in a short time.

3. Spaceborne Doppler Lidar

Figure 3 shows the schematic of laser Doppler measurements from space.

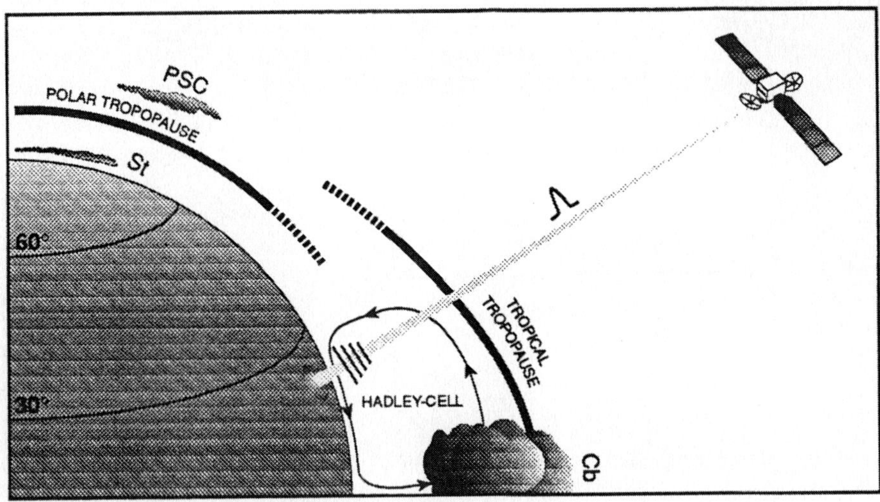

Fig. 3 Principle of a laser Doppler measurement

The Doppler lidar measures with an appropriate pulse repetition rate and a scanning mechanism (to cover a wide swath) a height resolved line-of-sight velocity component in the atmosphere. If one uses the laser Doppler system on board an aircraft or in space, the speed of the aircraft (spacecraft) modifies of the Doppler shift with

$$v_{LOS} = v_{LOS}(\text{wind}) + v_{LOS}(\text{carrier}) \qquad (3)$$

where $v_{LOS}(\text{carrier})$ is the carrier speed with respect to the laser line-of-sight.

Normally for ground based systems and airborne Doppler systems the second term $\frac{v_{LOS}^2}{c^2}$ in the Doppler shift is neglected. For a spaceborne unit with a carrier velocity of 7 km/s the second term produces a frequency shift in the same order as 1m/s wind velocity. Figure 4 shows the comparison of ground-based, airborne and spaceborne operations. The carrier velocity contribution is the main problem (figure 4, left side). It has to be corrected to get the wind component. The pointing accuracy gives the strongest requirement.

The wind field is assumed to be homogeneous in each level over the measured area. Scan patterns from a conically scanning system (figure 4, right side) show further problems. Starting from the sine-wave fit (eq.2) for the ground-based system one can perform a sector scan data processing for the airborne sensor. **But for the spaceborne sensor** the wind is not homogeneous over the wide swath. Therefore **only single LOS components can be handled.** Calculations of the horizontal wind out of two intersecting LOS measurements (two points are shown in figure 4, right side, lower part) are possible but not useful.

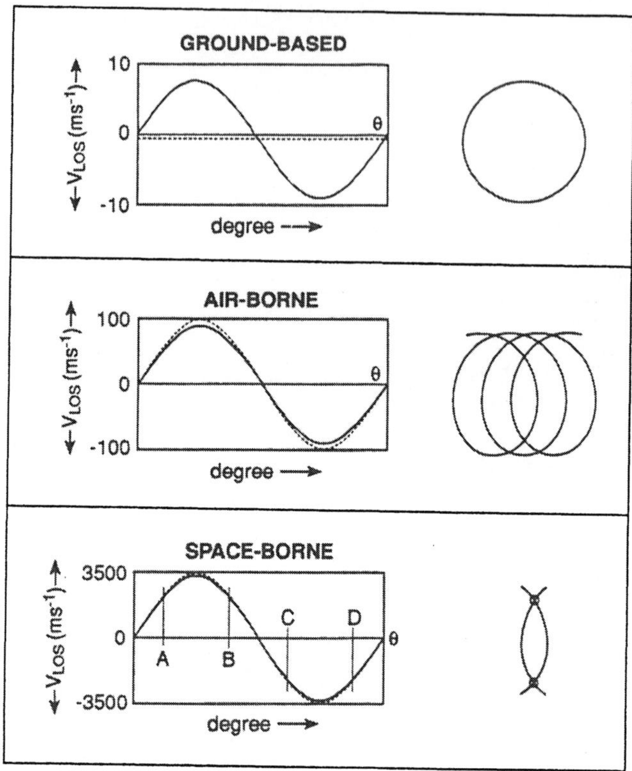

Fig 4 Comparison of ground-based, airborne and spaceborne wind measuring techniques

3.1 Representativity

The main question to be answered was: How representative is a single line-of-sight component ? It was assumed that a measured Doppler lidar wind profile (ground-based) from the sine-wave fitting was the true wind profile. A single LOS component out of this data set (points A to D in figure 4,lower left part) is the spaceborne lidar simulation value. The errors in measuring the LOS components with the Doppler lidar are mainly caused by speckle noise.

To improve the accuracy or probability for the estimation of the error, averaging is possible in various ways:

- integration of a few shots (by hardware before or by software after the signal processing)
- summing up some height levels (gates).

Data sets from a ground based system in the U.S. are available as single shots and as averages of three pulses (averages performed by hardware). Thus it is possible to compare both methods of averaging.

Figure 5 shows the influence of averaging on the error of the LOS component

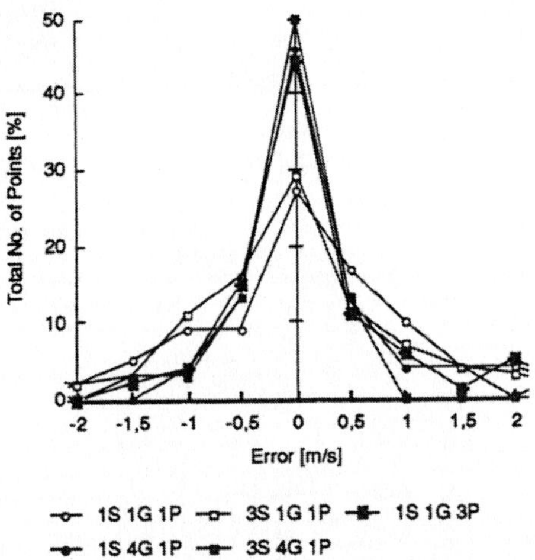

Fig. 5   Influence of method of averaging on the error of the LOS components.

The horizontal axis (error) is confined to ± 2 m/s , the abbrevations of the legend stand for:

1S: a measurement with single shots
3S: three shots averaged by hardware (receiver)
1G: calculations with the best range resolution (250 m)
4G: four range gates averaged (range resolution 1 km)
3P: software averaging of three shots (the selected angle and the two neighbour shots)

The results of the representativity study are (Werner et al.(1992)): The representativity depends mainly on the atmospheric aerosol loading and the laser stability. **Sufficient aerosols and laser pulse power produce a processable signal.** Averaging or/and selection of useful information out of a speckle signal is necessary.

3.2 System Simulation

The lidar simulation programme LIDAIR (Huffaker (1990)) is used to simulate the parameters of a proposed lidar system. (Figure 6) There are different input parameters:

> Platform (height is the starting level of the laser pulse, scan angle gives the LOS direction),
> Lidar parameter (laser, transceiver optics, detector)
> Windfield (a complete wind profile is included to be measured by the laser)
> Atmospheric modul: Hitran database, aerosol number density and backscatter profile, turbulence profile and atmospheric temperature and humidity are included to calculate the signal strength.
> Signal processing algorithms: Different processing algorithm (FFT, Poly Pulse Pair, Cramer-Rao etc. can be selected and finally horizontal wind calculations using VAD (aircraft) and the suspicious two-point-method (spaceborne) are possible.

The output of the simulation is a signal-to-noise (SNR) calculation for the different levels (fig.6). The signature SNR for a single LOS component shows the speckle effect.

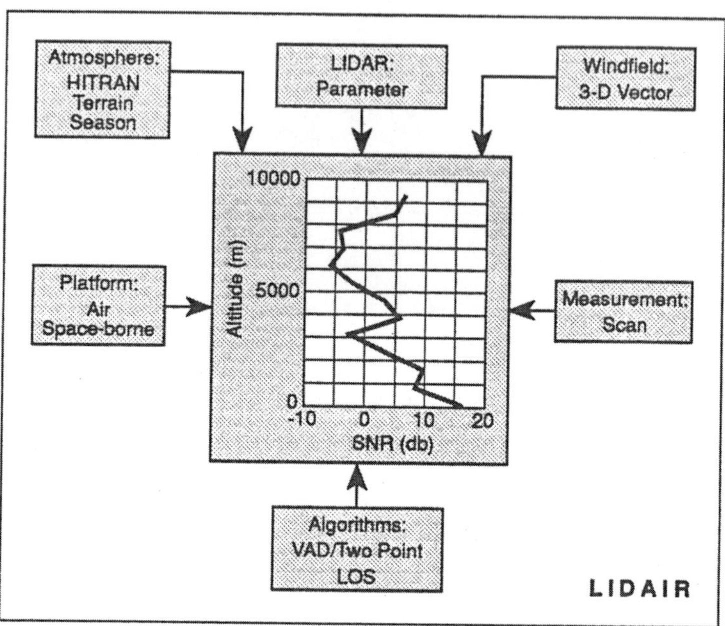

Fig. 6 Block diagramme of the system simulation

Losses of the system are summarised in table 2.
A CO_2 pulsed laser with 19 Ws pulse energy is used together with an 1.5 m diameter scanning telescope. The number of photons transmitted by the laser are $8.7 \cdot 10^{20}$ and 118 photons hit the detector from scattering back from an altitude between 7 and 8 km above ground. From this photons only 5 contribute to the radial wind estimate. For a given set of parameters (table 2) one can estimate the radial velocity error using the Poly-Pulse Pair processor. Figure 7 shows the error versus height.
The backscatter profile is the input together with the extinction and this gives an excellent wind velocity error in the lowest 6 range gates (1 km height resolution), gives good wind errors also up to the stratosphere and poor data above 20 km. Averaging over larger volumes (300 km * 300 km* 3 km) solves this problem.

Table 2 Parameters of a spaceborne Doppler lidar (Kavaya (1992))

PARAMETERS		
DOPPLER LIDAR		PHOTONS
Transmitted energy	19 Ws	
Photons		8.7×10^{20}
losses in transmitter optics	× 0.9	
losses through atmospheric extinction	× 0.71	
Photons hit the target		5.5×10^{20}
range R (orbit height 555 km, 45° cone)	776 km	
optics diameter D	1.5 m	
observation time	10.4 µs	
backscatter	$8.7 \times 10^{-11}\,\text{m}^{-1}\,\text{sr}^{-1}$	
Photons scattered back toward the mirror		220
losses through atmospheric extinction	× 0.71	
losses in the receiver optic	× 0.76	
Photons hit the detector		118
system efficiency	× 0.31	
detector heterodyne efficiency	× 0.4	
laser pulse spectrum loss	× 0.675	
unexplained coherent lidar loss	× 0.5	
Photons per radial wind estimate		5

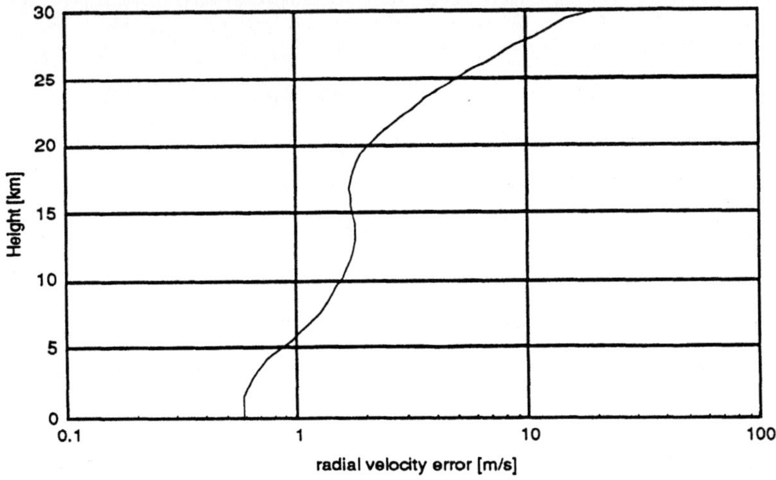

Fig. 7 Simulated radial velocity error vs. altitude for a spaceborne Doppler lidar

3.3 Wind field extraction from LOS measurements

Today the global models used for operational forecasting have horizontal grids of l degree in latitude and longitude compared with 3 to 4 degree ten years ago. The vertical resolution has increased simultaneously between 15 to 20 levels for vertical description.

To meet the requirements in the year 2000, the grid has to met the proposed observational requests. The frequency of observations and the coverage is another item. The frequency of observations is normally 4 times a day for the stratosphere and troposphere where as the surface has to be monitored 8 times a day.

There are a lot of discussions concerning the imortance of wind information from the tropical regions. It is essential for the objectives of international projects such as Gewex that the tropical regions are covered. Nevertheless for weather forecasting a global coverage is requested.

Any kind of wind information will be used in the forecasting offices in the year 2000. A sensor like ALADIN gives LOS components. These components can be used as input parameters as an ESA study started. Figure 8 shows the vertical resolution and the coverage by Doppler lidar observations. Region A is not affected by clouds where as region B is affected. In each volume of region A observations from different directions, different LOS components, are available.

These observations have errors and it is not useful to combine these data to a calculated horizontal wind using the 2-point-method (Representativity study Werner et al. 1992). The following procedure is proposed (figure 9). All the measurements in a certain level will be analyzed and filtered by a signal-to-noise estimator. Low S/N data and large backscatter data will be preselected. All other data points, LOS components, will be used as input data for the forecast. The angle of observation is an indicator for the wind field in the model. It could be possible that no data from a specific level can be transfered caused by cloud obstruction or by low signal to noise, for example in level k of region A or in levels 1 to m in region B.

The Doppler lidar acts as a backscatter lidar with lower range resolution. Also the information given by the backscatter lidar (cloud top height, height of the planetary boundary layer etc.) can be used to improve the radiation measurements.

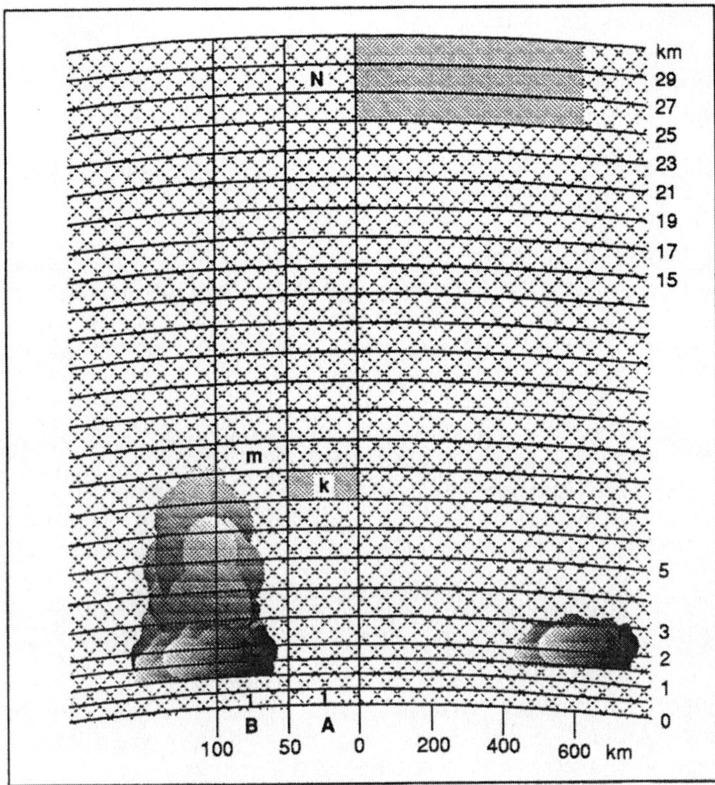

Fig. 8   Principle of spaceborne Doppler lidar measurements, grid size and levels

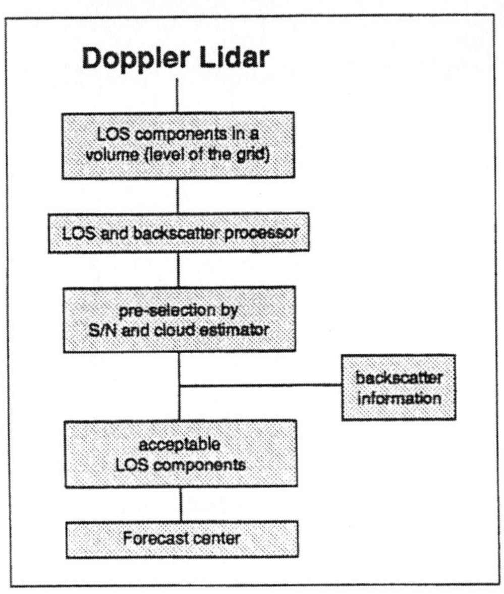

Fig. 9   Schematic of data handling

170

3.4 Critical Components

Beside the discussed technique of a single LOS measurement and the data handling, there are a few critical items before a system realisation:

Laser: A breadboard version of the proposed e-beam excited laser is developed.

Lag-angle compensation: A demonstrater on a mechanical design ist tested. Caused by the time difference between transmitting the laser pulse and receiving the backscattered signal, the spacecraft has moved and the pointing of the telescope has to be corrected (lag-angle).

Scanning: The total mass of the transceiver telescope has to be scanned. Ideas are requested to deal with the reduction of the rotating masses.

Pointing: Pointing is a major problem, especially if one would prefer shorter wavelengths instead of the 10 μm.

4. Summary

The paper showed the feassibility of a spaceborne Doppler lidar, The space agencies continue in the developmenmt. Critical components are tested and prototypes will be flown in aircrafts during the next years.
A platform for the spaceborne sensor is available with ALMAZ, the large Russian unmanned platform. A cooperation between DLR and the Institute of Atmospheric Optics of the Russian Academy of Sciences has started.

5. References

Hall,F.F.Jr.; R.M.Huffaker, R.M. Hardesty, M.E.Jackson, T.R. Lawrence, M.J.Post, R.A.Richter, and B.F.Weber: Wind Measurement Accuracy of the NOAA Pulsed Infrared Doppler Lidar. Appl. Opt. 23, 2503,1984

Huffaker, R.M.:LIDAIR, Coherent Technologies, P.O.Box 7488, Boulder,CO 80306,U.S.A.

Kavaya, M.J.: Projected Technical Specifications for a 2 mm Space-based Coherent Lidar Wind Sensing System. Proceedings 16th International Laser Radar Conference, Boston,MA July 1992- NASA Publication CP-3158

Köpp,F.;R.L.Schwiesow, and Ch.Werner: Remote Measurements of Boundary-Layer Wind Profiles Using a CW Doppler Lidar. Journ. Climate and Appl. Meteor. 23, 148-154 (1984)

Post M.J., Cupp R.E.: Optimizing a pulsed Doppler lidar; Appl. Opt. 29 (4145-4157) 1990.

Werner, Ch., J.Streicher, and G. Wildgruber: Representativity of Wind Measurements from Space. Proceedings of the Central Symposium of the International Space Year Conference, Munich, March 30-April 4 ,ESA SP-341, 1992 (This study was performed under ESA-contract No. 8664/90/HGE-1, the data for the simulation are from the NOAA-WPL-ground based Doppler Lidar.)

LIDAR-Visibility
Measurements
FOG

Visibility Measurements of Fog on Highways Using a Lidar Method in a Car

W. Hahn[1], W. Krichbaumer[2], J. Streicher[2], and Ch. Werner[2]

[1] BMW-AG, Knorrstr. 147, D - 80937 München
[2] German Aerospace Research Establishment (DLR), Postfach 1116, D - 82230 Wessling

Every year between fall and spring fog endangers the traffic on highways; serious accidents occur with a lot of cars involved. Government authorities regulate speed limits in relation to visibility, but nobody can determine the local visibility exactly.

A similar situation is well known from airport operation. Fog drastically reduces the landing capacity of the airports. Visibility at airports is measured continuously with transmissometers. A new sensor to get the slant visual range for the pilot is being introduced now at Hamburg airport. The basic element of this device is a modified laser cloud ceilometer (Impulsphysik).

It has been the intention of the authors to transfer the knowledge on the slant range visibility sensor to the highway situation. There are two approaches possible:

1) installation of visibility sensors on highways, which is already done at a few places;
2) installation in cars for control by the driver himself.

This paper reports on the development of a new sensor. Laser range-finders are currently installed in cars and trucks to measure the distance to a proceeding car (LEICA). A modification of such a sensor to measure visibility was made. A few problems had to be solved:

1) choice of wavelength with relation to the human eye for visibility measurements,
2) dependency of the wavelength on atmospheric turbidity,
3) laser eye-safety,
4) influence of multiple scattering at visibilities smaller than 200 m.

The wavelength used for the lidar sensor in the near infrared presents no real problem because the object to be sensed is fog appearing white which means that scattering from fog is wavelength-independent.

There are differences in the backscatter-to-extinction ratio for different fog situations. Advective fog and radiation fog are two different fog classes where simple backscatter sensors give different visual ranges, although the same visibility is measured by a transmissometer. It is also possible that other weather conditions (fog containing ice particles, snow etc.) may occur. The sensor must distinguish between these situations first. Polarisation sensitivity is one solution, multiple scattering is

174

the other.

The sensor is applied to small visibilities below 100 meters, where

a)   the signal is large and the eye-safety problem can be solved, and

b)   multiple scattering is dominant.

As known from airport operations of the laser ceilometer, one can use this multiple scattering contribution to determine the visibility.

Figure 1 shows a simulation of the sensor based on the DLR Lidar Simulation Programme (Werner et al. (1992); Streicher (1992)).

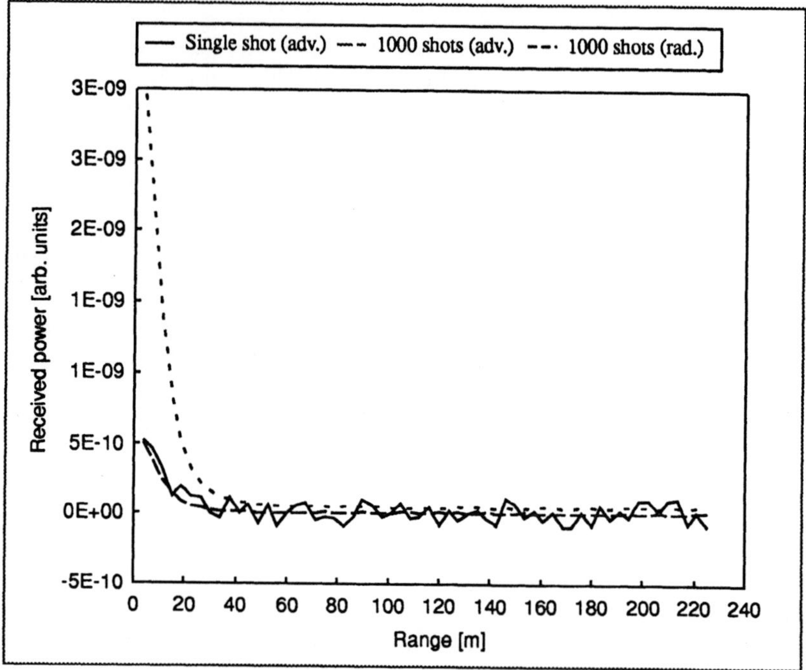

Fig. 1.: Simulated lidar signals for two fog situations with visibilities of 100 m

The solid curve shows a single shot with an eye-safe lidar system (output power about 0.8 µJ). The dashed curve is simulated under the same conditions as the solid one - advective fog with a homogeneous visibility of 100 m - but this time averaged over 1000 shots. The third plot (dotted) shows a lidar signal, again for 100 m visibility, but for radiative fog. A comparison of dotted curve and dashed curve shows that it is only possible to estimate visibility from backscatter if additional information on the atmospheric condition (fog type) is available because of the variable backscatter-to-extinction ratios for different types of fog: As is well known since Koschmieder's times, visibility is governed by extinction, not by backscatter.

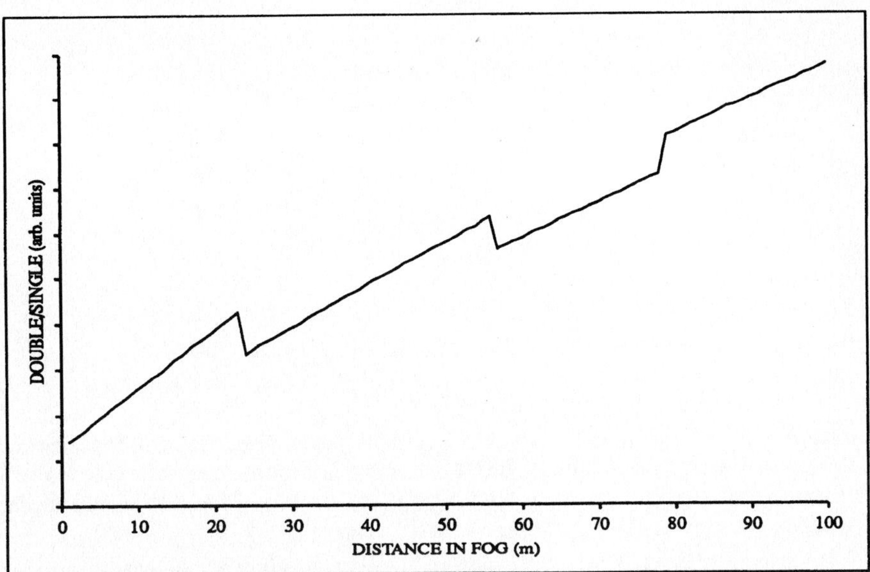

Fig.2.: Double scattering signal/single scattering signal vs. distance (simulated) from four homogeneous slabs of fog using different size distributions

The required additional information on the fog under consideration may be given by a sophisticated analysis of higher orders of scattering, i.e. if the assumption of single scattering as a basis of the evaluation of data is dropped (it is obvious from the high optical depths in fog situations that this assumption is wrong anyway) and if the lidar equipment is suitably modified to enable separate measurement of multiple scattering.

Figure 2 clearly indicates that multiple scattering in this case is indeed an additional source of information if analysed properly. The picture shows the result of a simulation using our stochastic model of multiple scattering (Krichbaumer and Oppel (1988); Oppel et al. (1989)). A monostatic coaxial lidar is assumed to measure in a fog atmosphere consisting of four homogeneous layers of different monodispersions of spherical particles (x=20, 15, 10, 5) but all with the same visibility (the same extinction). We have plotted the ratio double scattering signal/single scattering signal vs. distance from lidar. This ratio for each of these slabs is given approximately by (Γ is the field of view of the receiver; cf. Oppel et al.(1989), Ch. 7 and also Bruscaglioni et al. (1980))

$$\frac{\mu_2(r,\Gamma)}{\mu_1(r)} \approx A(\Gamma) \cdot \sigma$$

Therefore we should get for each slab a piecewise linear function of range for all four slabs. This is confirmed by the simulation in Fig. 2. Similar relations for higher orders of scattering exist.

Therefore, if multiple scattering information can be extracted properly, then the extinction coefficient (the visibility) can be estimated from the slope of the line plus additional information although different types of fog are present.

A prototype of such a sensor has been built; the method for the discrimination of the extinction has been tested. The ideas behind the system are in the stage of being patented; therefore no more detailed informations are given here.

Literature:

Bruscaglioni, P.; G. Milloni; and G. Zaccanti (1980): On the contribution of multiple scattering to Lidar returns from homogeneous fogs and its dependence on the Lidar range and on the receiver angular aperture, Opt. Acta **27**, 1229-1242.

Krichbaumer, W.; Oppel, U.G. (1988): A general stochastic model for simulation and calculation of multiple LIDAR backscattering and some reconstructions of scattering distributions from double scattering return signals. Proc 14th ILRC, Innichen-San Candido, Italy, June 20-24, 1988.

Oppel, U.G.; A. Findling; W. Krichbaumer; S. Krieglmeier; M. Noormohammadian (1989): "A Stochastic Model for the Calculation of Multiply Scattered LIDAR Returns", DLR-Report DLR-FB 89-36.

Streicher, J. (1992): Simulation of a Lidar to improve Visibility in Dense Fogs, paper accepted for presentation at SPIE conference "Lidar for Remote Sensing", Berlin, June 24-26, 1992.

Werner, Ch.; J, Streicher; H. Herrmann; H.-G. Dahn (1992): Multiple Scattering Lidar Experiments, accepted for publication in Opt. Eng.

Trial of an Eye-Safe Laser-Radar

J. Streicher
Institute of Optoelectronic
German Aerospace Research Establishment
P.O.B 1116
D-82230 Wessling

1. Introduction

An eye-safe laser radar has been developed from a cloud ceilometer in conjunction with HAGENUK GmbH, Hamburg. The system has a very low output power, a enlarged optic, and is therefore classified as eye-safe class 3A. A high pulse repetition rate and therefore the possibilty of averaging over some hundreds of shots compensates for the low output power. The device has been used so far for visibility measurements and for determining the cloud altitude [1].The range resolved detection of aerosols has been shown at an international experiment in England, fall 1992. The technical equipment and the installation of the complete system in a car will be described.
A further test, the measurement of industrial pollution, has been made at a chemical factory in Hamburg in early spring 1993. The correlation between the emission of acid loaded particles and specific phases of the production cycles will be presented.

2. Description of the laser radar system

A typical laser radar system consists of three modules: the transmitter of the light (a laser with optional optics), the receiver (a photodiode with an amplifier and again with optional optics) and the data aquisation (a fast analog/digital converter and some mass storage).
These three modules are, for the modified cloud ceilometer, integrated into two components:
- the laser radar system with transmitter, receiver, electronics, fast A/D converter, and a controlling unit (a 68000 computer)
- a Commodore Amiga 2000 computer, which is responsible for the mass storage, processing of the data, and an online color display.

A rough scetch of the system is shown in figure 1, the system specifications are listed below.

Laser:	GaAs laser diode array
Wavelength:	906nm
Pulse energy:	1.6µJ
Pulse duration:	50ns
Pulse repetition rate:	2.5kHz
Optical diameter:	14cm
Laser safety:	conforms to IEC 825/VDE 0837 class 3A
Measurement range:	from 3m to 1536m
Data acquisition and control:	Commodore Amiga computer via RS232

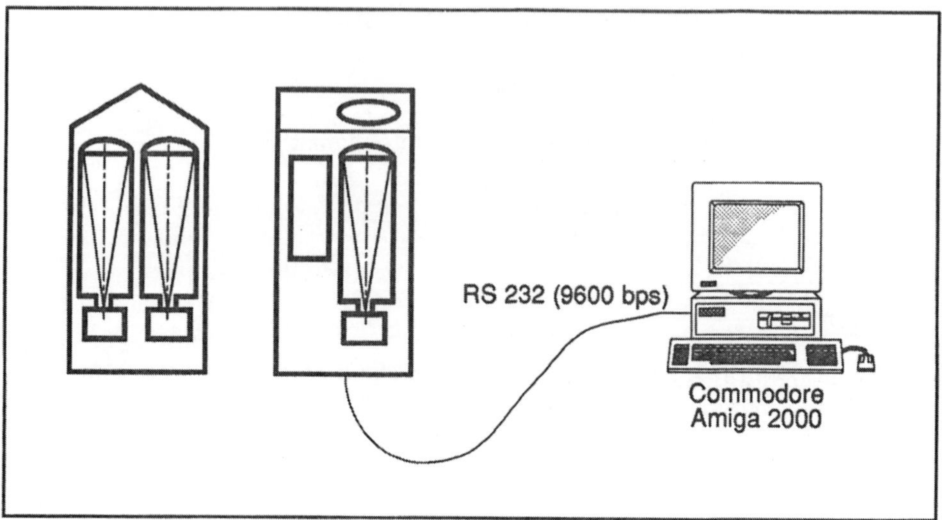

Figure 1: side and front view of the laser radar.

The laser matches the eye-safety criteria in two ways: it has the very low output power of 1.6µJ, and the laser beam is expanded to a diameter of 14cm. Due to the low output power the received signal is, for a single shot, highly contaminated with noise. Therefore one has to average over some hundreds of shots to get a representative information. This can be done very quickly with this system because of the high repetition rate of 2500 shots per second.

The averaging process has another positive effect, in addition to smoothing; the transient recorder (specially designed for this device) digitises every single shot with an 8bit A/D converter. The shots are then summed up in a 17bit sized array. The size of the digitising steps can be reduced from 1/256 (least significant bit of an 8bit converter) to 1/131072 (17bit), if the noise is greater than the least significant bit of the single-shot converter, i.e. noise level > 1/256. The total dynamic range of this fast (50MHz) transient recorder is therefore 17bit when averaging over many shots. This technique is called wide band noise dither [2].

3. Measurements of artificial fog diffusion

The participation in an aerosol diffusion experiment has been done in September 1992 with a trial in England.

The objective of this cooperative field study was to collect a comprehensive high-resolution meteorological and tracer data base in complex terrain using artificial fog and a trace gas (SF_6). Both were investigated by specialized high resolution sensors. One of these sensors was the DLR Lidar, the above described eye-safe infrared remote sensing laser radar.

The equipment for the measurement was installed in a car, the control unit (computer) at the back-seat (figure 2, top) and the measurement device (lidar) in the luggage boot (figure 2, bottom).

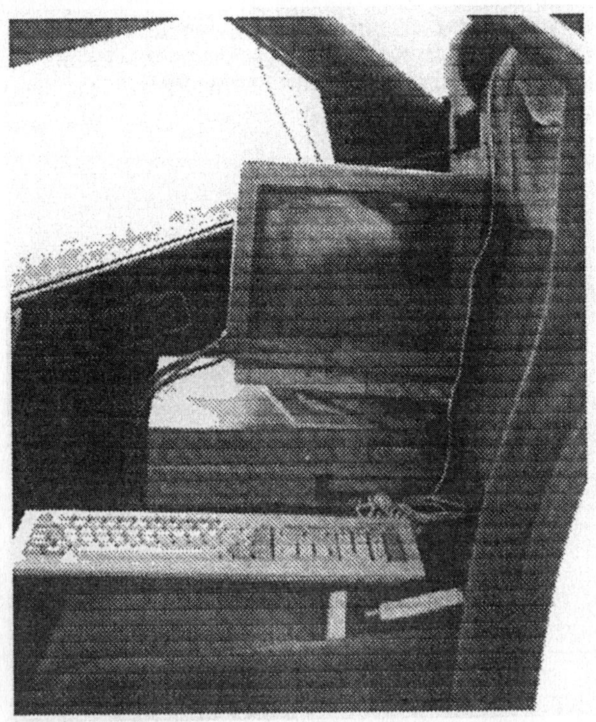

Figure 2: Installation of the laser radar system in a car

The startup time for a measurement is therefore very short (opening the luggage boot and power up the computer takes less than two minutes) and a measurement can even be done in moderate rain.
The placement of the lidar was always rectangular to the direction of propagation of

the plume, some hundred meters apart from the artificial fog source. The direction
and the elevation remained constant, so two-dimensional plots (mass concentration versus range) of the bypassing plume (versus time) have been taken. A time
series for a measurement in moderate rain is shown in figure 3.

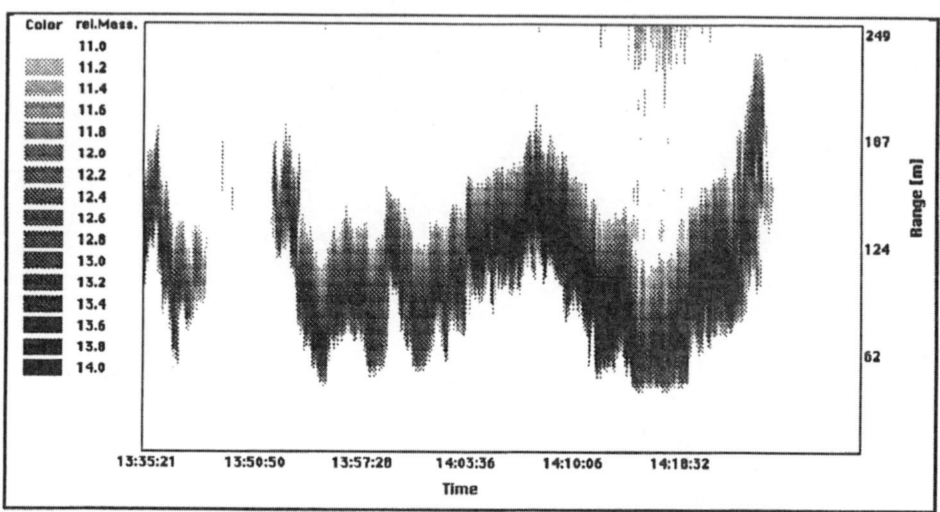

Figure 3: Artificial fog measurement in moderate rain (18. September).

The horizontal axis of these density plots represent the time at which the consecu-
tive measurements were made with a time resolution of three seconds, the vertical
axis the distance. The relative mass concentration is color coded, which means the
received intensity P(R) is range corrected by the square of the distance and applied
with the natural logarithm:

$$color = \ln(P(R)^*R^2)$$

The total mass concentration (in g/m3) can be found by taken further information
into account, like the particle distribution function (amount of particles versus particle diameter for the wavelength of the laser).
In the situation of moderate rain, as shown in figure 3, the plume stays almost compact and meanders only slightly (the gap between 13:40 and 13:50 was caused by
the shut down of the source). Figure 4 shows in the contrast a measurement with
no rain.

The two dotted lines were caused by system errors. In this case there are high fluctuations in the concentration as well as in the position of the plume (strong mean-
dering).
This picture reflects most of the measurements, which were taken in the complex
terrain over a time period of two weeks.
This shows, that the eye-safe laser radar is sensitive enough, to detect the small
particles (in the order of μm) of an artificial smoke plume, even in a range of some
hundreds of meters.

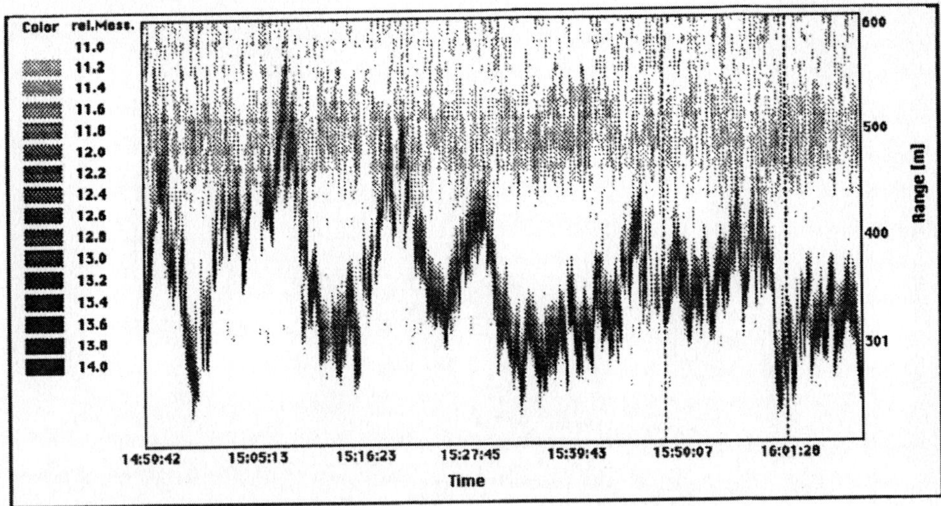

Figure 4: Artificial fog measurement, dry and windy (22. September).

4. Measurements of Industrial pollution

A more common application of a laser radar is the detection of industrial pollution. An additional test of the system has been made in spring 1993 at a chemical facto-ry in Hamburg. The problem at the factory was, that sometimes nearby parking cars had been damaged by acid loaded particles. The chimney, which emitted the parti-cles, was found quickly, by the position of the damaged cars, but the remaining questions were: when and why does the chimney release the particles ?
The installation of insitu sensors in 150 meters altitude was rejected, because of the high amount of technical and personnel costs. Therefore the optical remote sensing with a laser radar was choosen.
The system was placed on a observing platform in an altitude of 30m, the slant dis-tance to the chimney stack was constant at 465m, so the slant angle of the device was about 15 degree.
The data were stored on the hard disk of the computer, which capacity allowed a continued measurement over 24 hours. A backup on diskettes was made daily, to clear the hard disk for new measurements. The system worked automatically, only a few interrupts were observed: in the first night the computer had a breakdown, every morning when the sun light falled into the receiver the sun shutter closed, and for demonstration purposes.

The measurement began at the 1th of March at 15:30, a first incident can be clearly seen in figure 5 at 16:15.

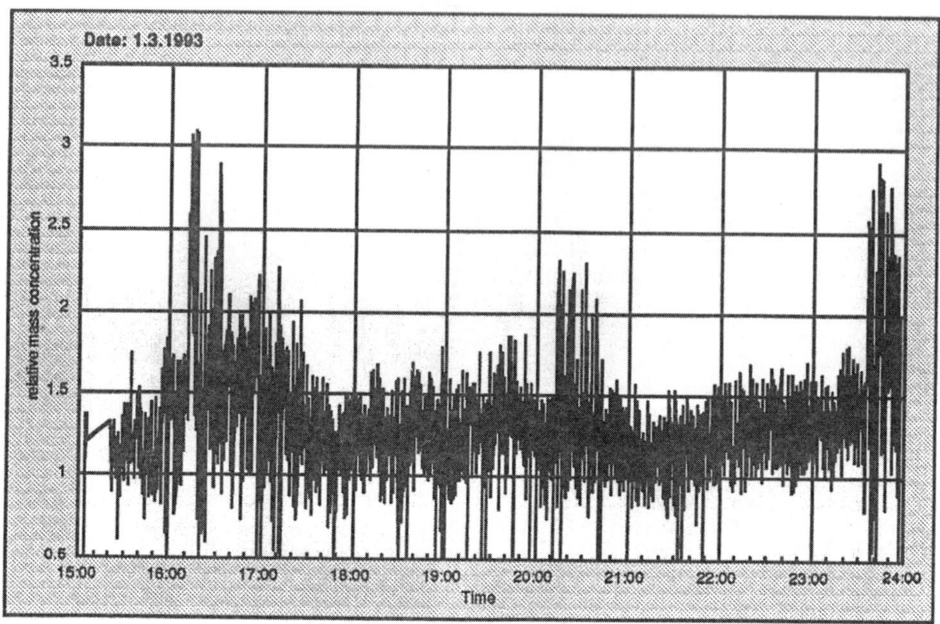

Figure 5: Relative mass concentration, at 465m averaged over 5 range bins

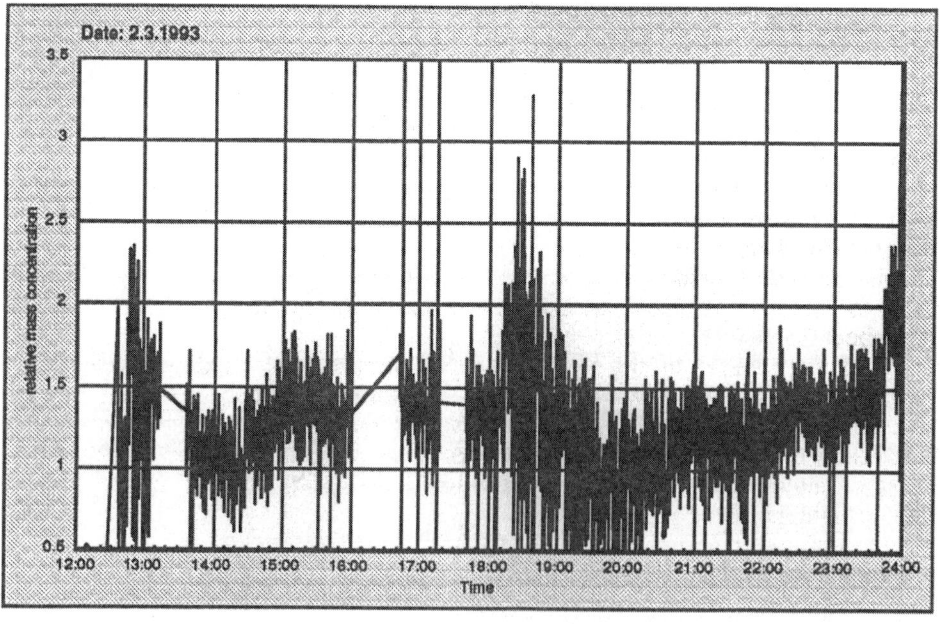

Figure 6: Relative mass concentration, at 465m averaged over 5 range bins

The comparison with the phase of the production cycle showed immediately, that the flow of mass of the plume had been reduced at that moment. Further investigations on the next day clarified this phenomena: a forced shutdown of one blower on the 2th of March at 12:50 showed again the increase of mass concentration in the plume (figure 6).

The other incidents at 18:15 and 23:40 can also be clearly related to decreases in the production cycle, caused by switching off one supply (the two peaks around 17:00 show the chimney as a hard target, the straight lines indicate no data available because saving was switched off).

It is planned to install a scanning laser radar system at the factory permanentely, for the observation of the complete area.

The measurement campaign was perfomed in conjunction with the engineering office E. Hansen, Bad Soden and Kayser-Threde, Munich.

5 Conclusions

The modified cloud ceilometer shows the following advantages in comparison to a traditional high power laser radar system:

- it is eye-safe and therefore anywhere installable, without restrictions
- the heigh repititionsrate, only available from a laser diode, allows high speed averaging, which increases the accuracy of the digitizer up to 17bits
- it is compact, therefore a quick installation in a car or an observation platform is possible
- it operates automatically, almost no operation control is necessary
- it is cheaper, the modifications, which lead to this prototype, will be included in the new generation of cloud ceilometers from Hagenuk GmbH

The measurements of artificial fog and at the chemical factory in Hamburg show the applicability of this eye-safe laser radar.

References

[1] Streicher, J., Münkel, C., Borchardt, H.
New slant visual range measuring device promises improved airport operations
ICAO Journal Vol 47, No. 12 pp 14-16 (1992)
[2] Vanderkooy, J., Lipshitz, S.P.
Resolution below the least significant bit in digital systems with dither
J. Audio Eng. Soc., Vol 32, No. 3 pp106-112 (1984)

Near-Field Effects in a Monostatic Multiple-Aperture Lidar

V. Geinitz, W. Richter
TU Ilmenau, Fakultät für Maschinenbau, Postfach 372, D - 98693 Ilmenau

W. Krichbaumer
DLR, Institut für Optoelektronik, Postfach 1116, D - 82230 Wessling

For many years backscatter lidars have been used for remote sensing of the atmosphere. Although the determination of atmospheric parameters in thin aerosol by lidar is relatively easy, special care has to be applied when the lidar is used in dense atmosphere. The reason for that is that in clouds or fog one has to be aware of multiple scattering, whereas the usual evaluation procedures for the signals take only single scattering into account. Therefore, DLR together with Munich University developed a stochastic model for the mathematical treatment of multiple scattering(/1/,/2/). Evaluation of the contribution by multiple scattering to the signal provides additional information on the scattering particles. To this end modification of the lidar hardware is necessary with respect to the possibility of a separation of single and multiple scattering contribution as far as possible; moreover, simultaneous measurement of polarisation is indispensable. The DLR-Microlidar meets these requirements: The receiver is separated into a central part for the detection of (mainly) single scattering radiation and in a ring-shaped outer part for the detection of (mainly) multiple scattering contributions. For each of the two regions the received radiation is split up into the parallel and orthogonal polarisation components.

A problem for measurements in a cloud within a short distance from the telescope is oversaturation of the detectors. By use of a diode-pumped laser (/3/) the output power can be reduced by tuning down the pump diode to prevent oversaturation. This allowed also near-field measurements. This produced a new effect: Up to now we assumed that the ring-shaped detector would receive only multiple scattering contributions to the signal and no single scattering, but in case of near-field measurements we found in a small distance (some metres) from the telescope a signal in the "multiple scattering channel" whereas no signal was registered in the "single scattering channel" (Fig. 1).

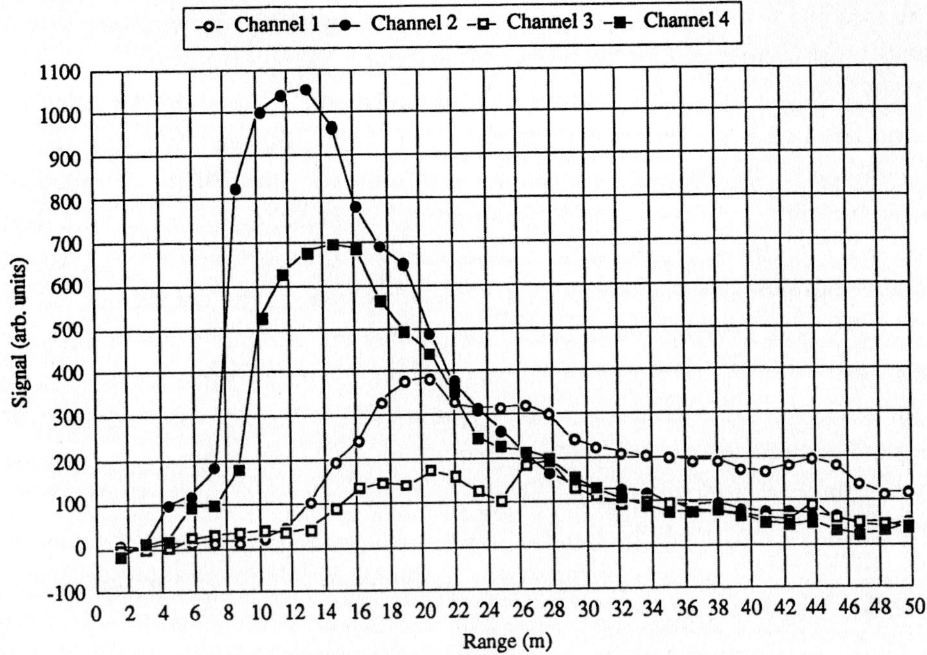

Fig. 1: Near-field signal of the Microlidar. Channels 1 and 3: central detector; Channels 2 and 4: ring-shaped detector

From path length and polarisation it was concluded that this signal could have been produced only by directly backscattered light. How is that possible?

Fig. 2 shows the mirror system, the laser, the laser beam with divergence Θ, and the limiting rays for the central and the ring-shaped detectors.

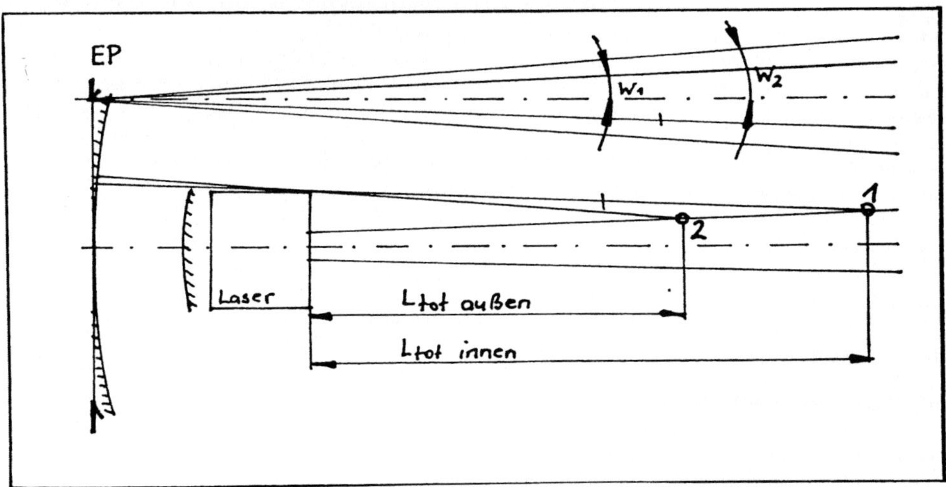

Fig. 2: Optical configuration of the Microlidar telescope

The rays incident on the telescope under the limiting field of view angles w1 and w2 can just be received by the interior or exterior telescope. They were drawn in such a way that they hit the main mirror exactly along the central obstruction region. The points of intersection of these rays with the laser beam mark the "blind" region for the interior and for the exterior detector. Obviously radiation is received first by the exterior detector (point 2 of Fig. 2, at about. 4m) and then from a greater distance also by the interior (point 1, at about. 9m).

Many parameters define the radiant power received by the detectors: field of view, size of the mirror system and of the diaphragm, set up and size of the laser unit, beam profile and divergence. Based on photometric-optical considerations a model for the Microlidar has been developed which describes its function, taking into account the geometry of the lidar. Here the detectors were placed in the focal plane of the mirror system.

The set up of the coaxial monostatic lidar was taken to be rotational symmetric. This applies also to the laser for which a circular housing was assumed which produces the central obstruction.

The beam profile in the model has TEM00 characteristics and is described mathematically by a Gaussian profile. Lasers with arbitrary modes can be taken into account expressing the size of the beam waist in terms of the fundamental mode and applying a correction term (/4/). The laser in the model is cw, contrary to the laser used in reality.

It was assumed that the atmosphere is homogeneous and has no influence on the laser radiation. This simplification is justified because a model for the effects of the atmosphere (/1/,/2/) already is available and both models can be linked together.

A dense cloud or fog is simplified to a screen placed in a distance z in front of the receiver. The laser gives rise to a certain irradiance on the screen which via Lambert's law is transformed into a radiance constant in all directions of the half space. It is assumed that no radiation penetrates into the screen but that only diffuse reflection occurs.

The cross-over function of the transmitter and receiver cones is determined by calculating how much of the radiation reflected by the screen hits the telescope under a certain angle w.

Figure 3 shows the hitherto modelled lidar and the screen.

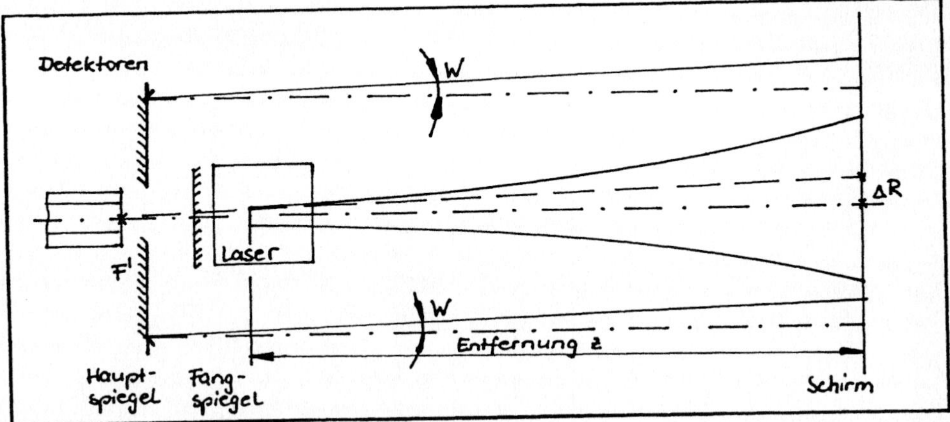

Fig. 3:   Lidar model and screen

Irradiance and radiance on the screen are - same as the laser profile - distributed symmetrically around the optical axis. The rays incident with angle w were drawn in the margin of the entrance pupil and represent the field of view of the telescope. On the screen, the centre of the field of view is translated with respect to the centre of the laser beam by ΔR. Therefore also the radiation intensity incident on the telescope is no longer symmetric with respect to the centre of the telescope. As a simplification, an average of the radiance was used in the model.

On the detectors, this averaged radiance gives rise to an averaged irradiance. However, caused by the central obstruction not all of the radiation hits the telescope.

The averaged irradiance on the detector area is transformed into radiant power which is converted into an electric signal.

Using the given parameters a formula for the radiant flux Φ was created which only depends on the distance z and the detector under consideration (interior or exterior).

The resulting curves are, after having reached a maximum, continuously descending. It is obvious that first the exterior detector receives a very strong signal from short distance. From the interior region only in greater distance reflected rays can be received. Thus the curves confirm the near-field effect. Qualitatively they are in good agreement with the measured curves.

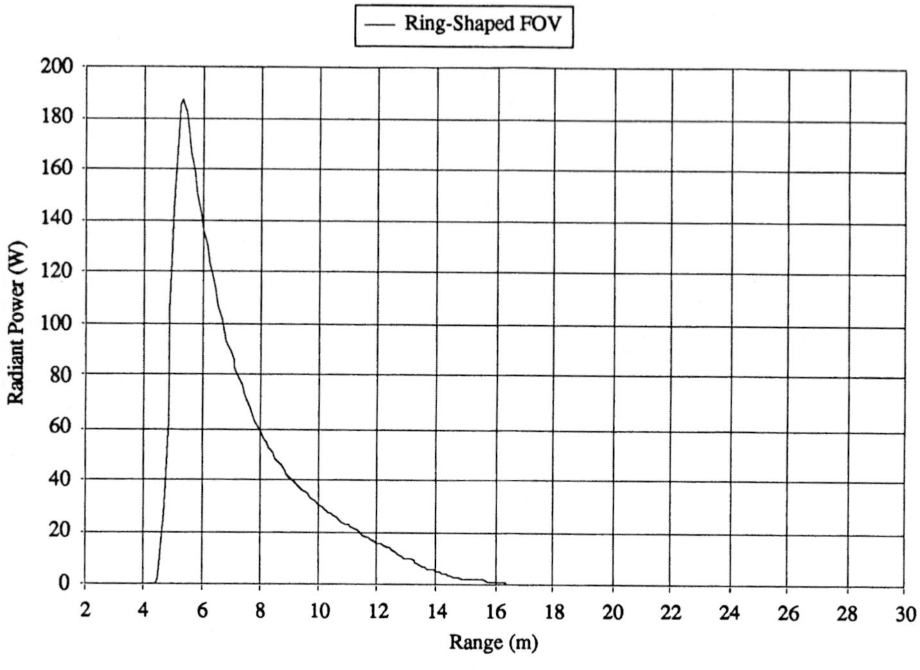

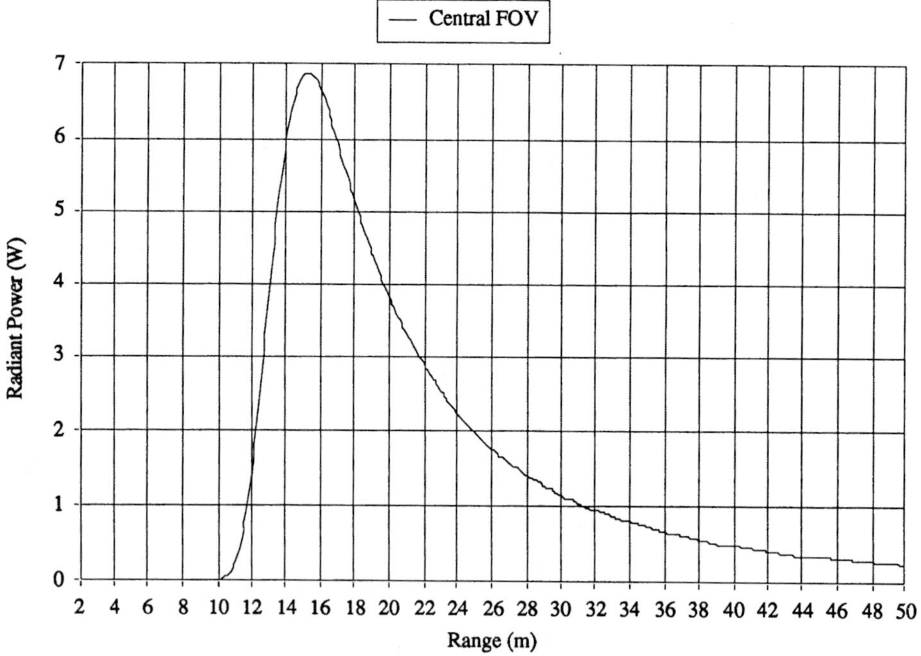

Fig. 4: Radiant power for the exterior (upper curve) and the interior (lower curve) detector

Literature:

/1/   Oppel, U.G., and Krichbaumer, W.: "Calculations of multiply scattered lidar return signals from slabs of dense aerosols", Proc. MUSCLE 4, 4th International Workshop on Multiple Scattering Lidar Experiments, Florence (Italy), Oct. 29-31, 1990.

/2/   Krichbaumer, W., and Oppel, U.G.: "Relation of multiply scattered return signals to the interpretation of airborne measurements with the DLR-Microlidar", Proc. MUSCLE 4, 4th International Workshop on Multiple Scattering Lidar Experiments, Florence (Italy), Oct. 29-31, 1990.

/3/   W. Krichbaumer , H. Herrmann, E. Nagel, R. Häring, J. Streicher, Ch. Werner, A.Mehnert, Th. Halldorsson, S. Heinemann, P. Peuser, N.P. Schmitt: "Diode-Pumped Nd:YAG Lidar for Airborne Cloud Measurements" , accepted for publication in: Optics & Laser Technology.

/4/   Ewert,T.; Richter,W.: "Laserprofiltransformation - analytische Schreibweisen und Analogien zur geometrisch - optischen Abbildung", Wiss.Z. TH Ilmenau 37 (1991) Heft 2.

Expertensysteme zur Datenanalyse technischer und ökologischer Multisensorsysteme

M. Noormohammadian, U.G. Oppel, A.V. Starkov*

Mathematisches Institut der Ludwig-Maximilians Universität München
Theresienstr. 39, 80333 München, Germany

* Computing Center, Siberian Division, Russian Academy of Sciences
Pr. Acad. Lavrentieva, 6, 630090, Novosibirsk, Russia
(z. Zt am Mathematischen Institut der LMU München)

Abstract

For the evaluation of a multisensor system we design a prototype of an expert system which is based on a causal probabilistic network. Using stochastic cause-effect relations, with such an expert system it is possible to reduce uncertainty about the evaluation of some sensors by introducing evidence from other sensors. In traffic guidance systems a LIDAR sensor may be used to detect obstacles and to determine visibility. We show how the visibility may be obtained from the LIDAR signal using additional knowledge from the other sensors and the inference system of the expert system.

1. Einleitung

Biologische Systeme sind seit jeher in der Lage in einer höchst komplexen und teilweise gefährlichen Umwelt zu existieren, weil sie über ein raffiniertes und im Laufe der Evolution "optimal" an ihre Lebensräume angepaßtes System von vielen Sensoren verfügen. Aber nicht allein diese Sensoren machen das Zurechtfinden in der Umwelt möglich, sondern es sind in noch höherem Maße die in diesen biologischen Systemen integrierten Auswertungsverfahren. Diese werten mit den Sensoren erfaßte Umweltdaten zusammen mit bereits gemachten Erfahrungen, Erkennungsvorgängen und Verhaltensregeln aus.

Zusätzlich ist es dem Menschen durch Anwendung naturwissenschaftlicher Erkenntnisse und Methoden gelungen, künstliche Sensoren zu entwickeln, die in speziellen Situationen alle bisher bekannten natürlichen Sensoren übertreffen. Zur Auswertung der Meßergebnisse künstlicher Sensoren bedarf es zum einen einer Theorie zur Entwicklung von Auswertungsverfahren und zum anderen der Erfahrung zur Beurteilung, ob die notwendigen Hypothesen für die Anwendung der Theorie erfüllt sind.

Erst der durch die rasche Miniaturisierung der Sensoren ermöglichte gleichzeitige Einsatz mehrerer Sensoren zur Erfassung einer Umweltsituation bringt zusätzliche Informationen und hilft damit, falsche Auswertungen der Meßergebnisse einzelner Sensoren zu reduzieren. Die gleichzeitige oder iterative Auswertung eines solchen Multisensorsystems erlaubt die Steigerung des Vertrauens in die Ergebnisse der Auswertung durch Anwendung von Regeln, die aus der Erfahrung des meist stochastischen Zusammenhangs der Meßergebnisse der verschiedenen Sensoren abgeleitet werden können.

Diese Erfahrungen können auf statistisch gewonnenem Wissen oder auf Wissen von Experten basieren. Ihre Auswertung wiederum kann durch Experten und durch Methoden der künstlichen Intelligenz geschehen. Insbesondere bei großen Multisensorsystemen, bei massenweise anfallenden Daten oder hohen Anforderungen an die Geschwindigkeit der Auswertung werden computergestützte Methoden der künstlichen Intelligenz notwendig.

Solche Multisensorsysteme und Auswertungsmethoden spielen bei ökologisch-meteorologischen Anwendungen (z.B. bei boden-, luft- oder raumgestützten Systemen zur Erfassung von Aerosoldaten für Imissionskontrolle und Klimamodelle) und verkehrstechnischen Anwendungen (boden- oder fahrzeuggestützten Systemen zur Ermittlung von Sichtweiten und Hindernissen; vgl. [1], [2] und [3]) eine wichtige Rolle. Diese Expertensysteme, deren Wissensbasis teilweise unsicheres Wissen enthält, sind teils regelbasiert und teils stochastisch. Die Wissensbasis stochastischer Systeme ist in Form kausal-probabilistischer Netze organisiert und kann mit der Shell HUGIN konstruiert und operiert werden. (vgl. [4] und [5])

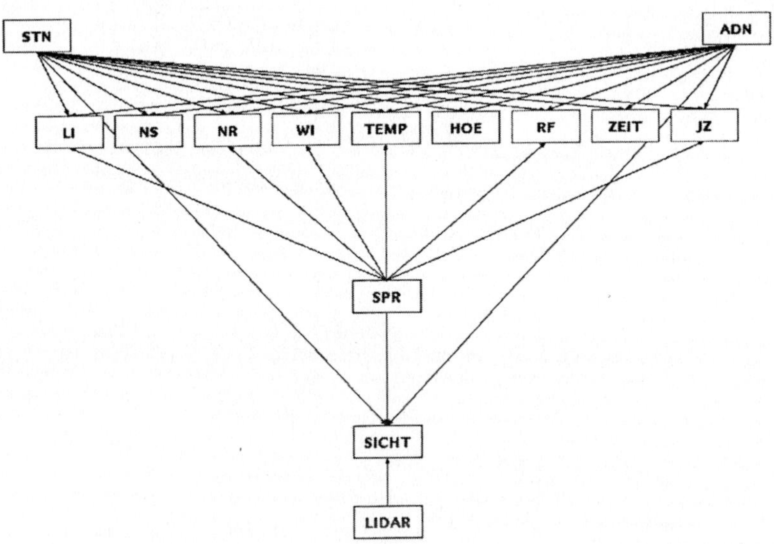

Bild 1: Beispiel eines mit HUGIN erstellten kausal-probabilistischen Netzes für ein fahrzeuggestütztes Multisensorsystem (siehe Abschitt 3)

2. Grundprinzipien kausal-probabilistischer Expertensysteme

Wir wollen dies im folgenden kurz für die Auswertung der Meßdaten eines LIDAR-Sensors im Rahmen eines Multisensorsystems darstellen. Im nächsten Abschnitt werden wir diese Gedanken an einem stark vereinfachten Beispiel konkretisieren.

Bei der Auswertung des LIDAR-Sensors treten umweltbedingte Unsicherheiten auf. Diese werden mit Hilfe teilweise ebenfalls unsicherer Daten anderer Sensoren schrittweise verringert und bezüglich ihrer verbleibenden Unsicherheit neu bewertet. Dieser Prozeß der Neubewertung erfolgt z.B. bei der Shell HUGIN automatisch durch Iterationen von kausal-probabilistischen Schlüssen nach dem Bayesschen Prinzip.

LIDAR-Signale werden auf der Basis von Gleichungen ausgewertet, die wegen der Komplexität der Umwelt immer mehrere Umweltparameter enthalten. Dies gilt für die üblicherweise angewandte und nur den einfach gestreuten Anteil des Signals erfassende klassische LIDAR-Gleichung; sie enthält stets teilweise oder ganz unbekannte Parameter wie den Rückstreukoeffizienten und den Extinktionskoeffizienten.

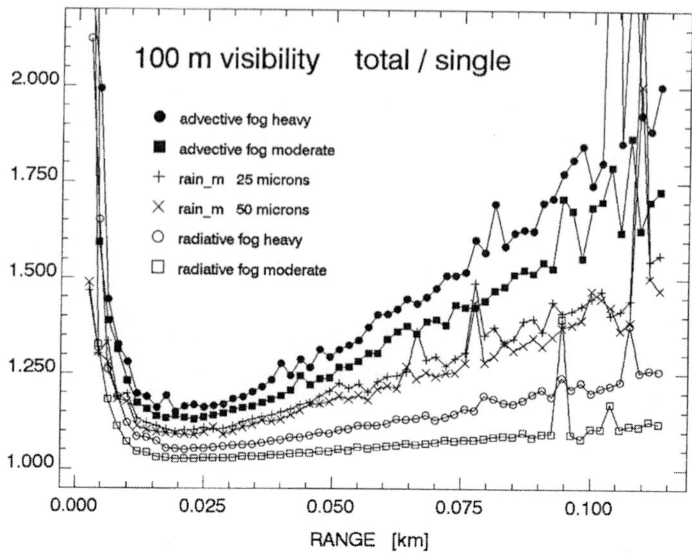

Bild 2: Verhältnis des mehrfach zum einfach gestreuten Anteil des LIDAR Rücksignals
(Monte Carlo Simulation)

Bei dichten Aerosolen tritt in erheblichem Maße Mehrfachstreuung auf, welche mit Hilfe transport-
theoretischer Methoden (vgl. [6]) oder stochastischer Prozesse (vgl. [7] und [8]) modelliert werden
kann. Das LIDAR-Rücksignal aus dichten Aerosolen kann nur auf der Basis der so gewonnenen exakten LIDAR-Gleichungen ausgewertet werden, da sie auch die höheren Streuordnungen erfassen.
Natürlich enthalten diese exakten LIDAR-Gleichungen viele unbekannte Umweltparameter. Neben
dem örtlich meist variablen Extinktionskoeffizienten sind dies beispielsweise die vom Streuort und
von der Streurichtung abhängigen (und durch die Phasenfunktion des streuenden Partikels beschrie-
benen) Streuwahrscheinlichkeiten.

Es geht nun darum, durch Beschaffung von möglichst viel Zusatzinformationen über einige dieser un-
bekannten Parameter möglichst sicheres Wissen über die anderen gesuchten Parameter zu gewinnen.
Dieses Wissen über die gesuchten Parameter kann durch simples Vergleichen von numerisch simulierten oder in (weitgehend) bekannten Situationen gemessenen Signalen mit den zu analysierenden
Signalen oder durch Anwendung geeigneter Inversionsverfahren gewonnen werden. Da das Wissen
über einige der Parameter unsicher ist, ist auch das Endergebnis unsicher.

Es gilt, den Grad dieser Unsicherheit der bei der Analyse verwendeten Parameter und des Endergeb-
nisses abzuschätzen und ihn durch geignete Verfahren zu reduzieren. Dies kann durch die Verwendung
von Datenbanken, automatisierten Vergleichsverfahren und teilweise formalisiertem Expertenwissen
geschehen. Programme zur Strukturierung des Wissens und der Automatisierung oder Unterstützung
von Entscheidungs- und Bewertungsverfahren nennt man Shell - sie dienen zur Konstruktion von Ex-
pertensystemen.

Ein Expertensystem besteht aus einer Wissensbasis (sie enthält Fakten und Regeln), einer Pro-
blemlösungskomponente (ihr Inferenzmechanismus leitet aus Regeln und Fakten neue Aussagen ab),
einer Dialogkomponente (sie erlaubt dem Benutzer das Eingeben von Fakten und Regeln), einer
Erklärungskomponente (sie soll die Schlußweise des Systems nachvollziehbar machen durch Erläute-

rung der verwandten Fakten und Regeln) und einer Wissenserwerbskomponente (sie unterstützt den Aufbau der Wissensbasis durch Redundanz-, Konsistenz- und Regelüberprüfung).

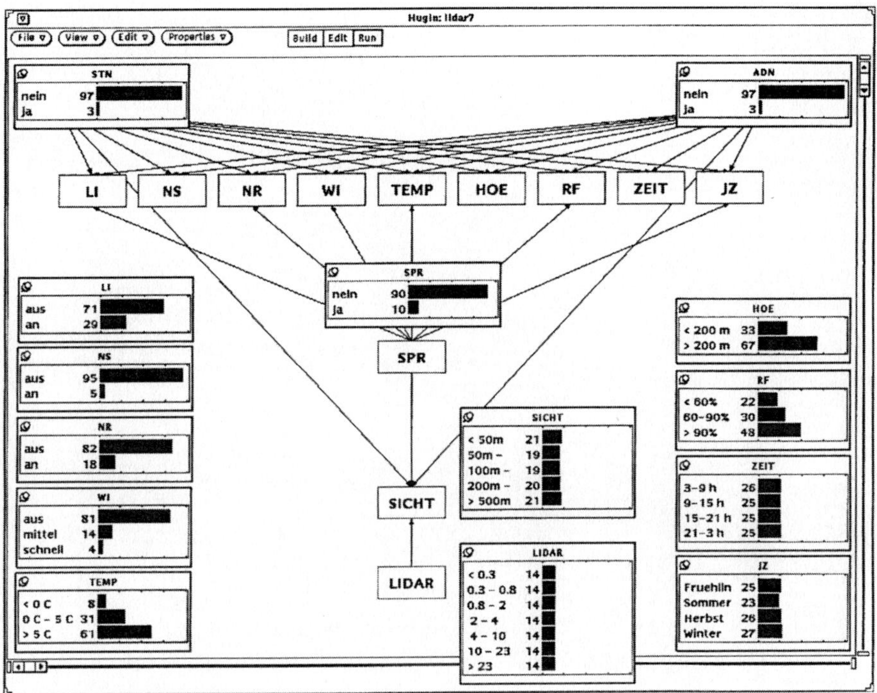

Bild 3: Das Gesamtwissen des kausal-probabilistischen Netzes in Form von Wahrscheinlichkeiten in den Knoten, welches aufgrund des globalen strukturellen Grundwissens (in Form der Netzstruktur) und des lokalen quantitativen Grundwissens (in Form von Tabellen bedingter Wahrscheinlichkeiten) vorgegeben ist.

Enthält die Wissensbasis unsicheres Wissen und nichtdeterministische Regeln, so ist es zweckmäßig die Wissensbasis in Form eines kausal-probabilistischen Netzes zu organisieren. Ein kausal-probabilistisches Netz kann mit der Shell HUGIN konstruiert und operiert werden. Es besteht aus Knoten und gerichteten Kanten (Pfeilen). Die Knoten beschreiben die wegen der vorhandenen Unsicherheit stochastischen Zustände des Systems (z.B. anfänglich gesetzte Umweltparameter, Ergebnisse von Auswertungsverfahren, iterativ verbesserte Umweltparameter, Expertenschätzung). Die gerichteten Kanten beschreiben qualitativ die stochastischen Relationen dieser Zustände. Quantitativ werden diese Relationen durch Tabellen bedingter Wahrscheinlichkeiten beschrieben (z.B.: in einer gegebenen Höhe und in einem gewissen Temperaturbereich sind aus H_2O bestehende Aerosole mit einer Wahrscheinlichkeit von 80% eisig). Diese bedingten Wahrscheinlichkeiten stellen zusammen mit der Netzstruktur (Knoten, gerichtete Kanten) das Expertenwissen dar. Die Bewertung der Zustände geschieht durch Angabe ihrer Wahrscheinlichkeitsverteilungen, die durch die Shell HUGIN aus dem anfänglichen Expertenwissen bestimmt werden.

Bei Eingabe neuer Erkenntnisse (z.B. Zusatzinformationen weiterer Sensoren, von einem Experten oder einer überschlägigen Rechnung) bewertet HUGIN die Zustände neu, und zwar im Einklang mit dem alten Wissen und den neuen Erkenntnissen. Insbesondere sind kausal-probabilistische Expertensysteme wie HUGIN für solche Analysesysteme geeignet, die aufgrund neuer Erkenntnisse iterativ neu bewertet werden müssen.

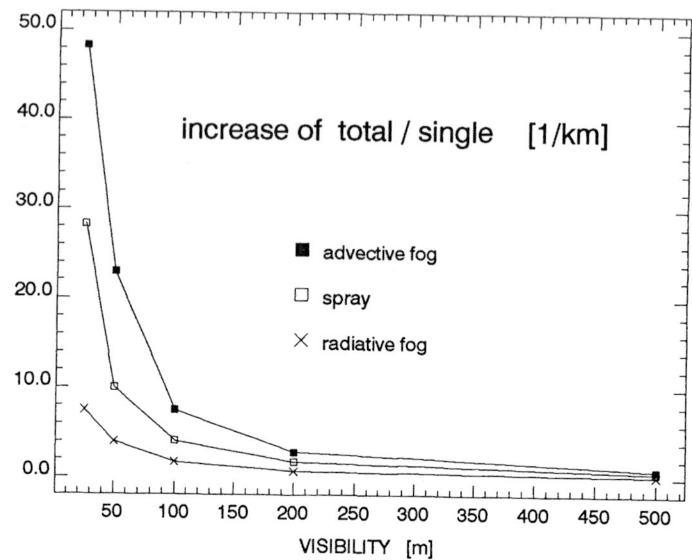

Bild 4: Steigung des Verhältnisses des mehrfach gestreuten Signalanteils zum einfach gestreuten Signalanteil in Abhängigkeit der Normsichtweite

3. Beispiel eines einfachen Multisensorsystems zur Sichtweitenbestimmung

Wir betrachten nun ein auf ein Fahrzeug montiertes monostatisches LIDAR mit dem (unter anderem) die Sichtweite bestimmt werden soll. Dieses LIDAR habe (wie das Micro-LIDAR der DLR) einen kollimierten und zum konischen Blickfeld des Empfängers koaxialen Sendestrahl. Der innerste Teil des Blickfeldes kann durch eine ringförmige Lochblende ausgeblendet werden; in dieser Konfiguration wird kein einfach gestreutes Rücksignal registriert. Außerdem läßt sich der äußere Teil des Blickfeldes durch eine Lochblende ausblenden; in dieser Konfiguration wird sowohl einfach als auch mehrfach gestreutes Rücksignal registriert. Aufgrund theoretischer Überlegungen kann man aber aus diesen beiden Signalen den einfach gestreuten Anteil vom mehrfach gestreuten Anteil mit einem Rechentrick trennen. Wiederum aufgrund theoretischer Überlegungen läßt sich aus dem Verhältnis des mehrfach gestreuten zum einfach gestreuten Signalanteil der Extinktionskoeffizient und damit die Sichtweite (SICHT) bestimmen. (vgl. [7]) Diese theoretischen Ergebnisse sind durch Berechnungen und Messungen bestätigt.

Der Extinktionskoeffizient ist proportional zum Verhältnis des mehrfach zum einfach gestreuten Signal. Der Proportionalitätsfaktor (LIDAR) hängt im wesentlichen nur von der Klasse der vorliegenden Streuer ab. Für die verschiedenen Klassen der Streuer (vgl. [9] und [10]) des Sprays (SPR), des advektiven Nebels (ADN) und des Strahlungsnebels (STN) sind diese Proportionalitätskonstanten verschieden und können im vorhinein berechnet werden. (z.B. mittels entsprechender Monte Carlo Simulationen wie in Bild 2; vgl. [8])

Um Sichtweiten zu bestimmen, braucht man darum nur dieses Verhältnis und die Klasse der Streuer zu kennen (vgl. Bild 4 und auch [11]). Informationen über die Klasse der Streuer aber können mit

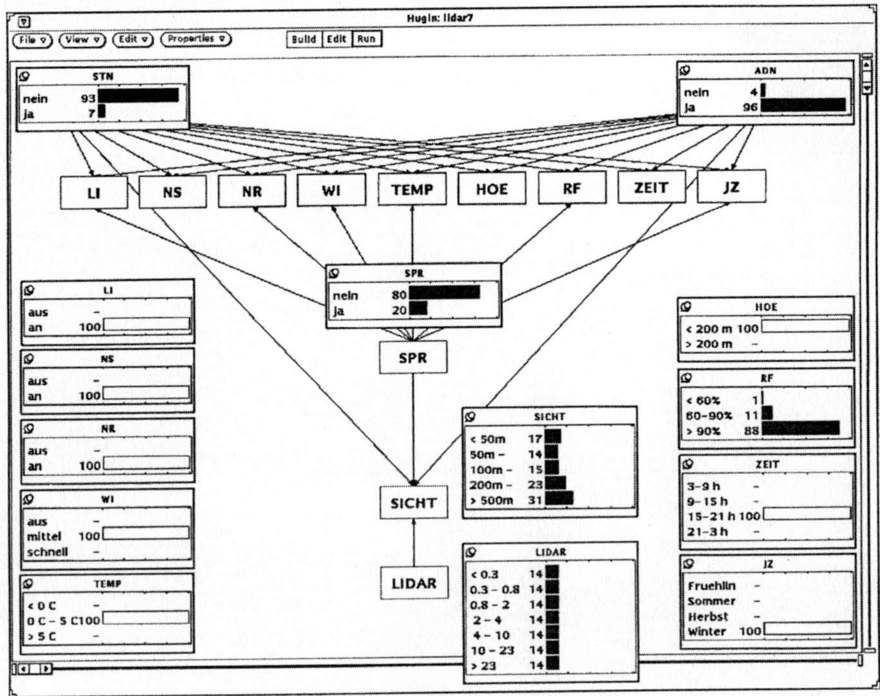

Bild 5: Die Eingabe von neuem Wissen wie "Frontscheinwerfer an", "das Fahrzeug befindet sich weniger als 200 m über dem Meeresspiegel (an der Küste ?!)", "es ist ein Winterabend", "die Außentemperatur liegt knapp über 0° C", etc. führen nach einer Neubewertung in Übereinstimmung mit dem alten Grundwissen zu einer *neuen* Wahrscheinlichkeitsbewertung. Die Wahrscheinlichkeit für das Vorliegen eines advektiven Nebels unter diesen Umständen ist deutlich erhöht.

Hilfe eines kausal-probabilistischen Expertensystems und einigen zusätzlichen Sensoren gewonnen werden. Wir verwenden in unserem Beispiel Indikatoren für die Beleuchtung (Frontscheinwerfer (LI), Nebelscheinwerfer (NS), Nebelrücklicht (NR)), Scheibenwischer (WI), die Temperatur (TEMP), die barometrisch gemessene Höhe (HOE), die relative Feuchte (RF), sowie die Tages- (ZEIT) und Jahreszeit (JZ). Wir nutzen hier also sowohl physikalische Sensoren als auch (indirekt) menschliche Sensoren. Mit Hilfe des in Bild 1 angegebenen kausal-probabilistischen Netzes, des darin in Form von bedingten Wahrscheinlichkeiten eingegebenen Expertenwissens (vgl. Bild 3) und der durch die zusätzlichen Sensoren gelieferten Evidenz wird mit Hilfe der Shell HUGIN eine neue, für die momentane Umweltsituation des Kraftfahrzeuges zutreffende Wahrscheinlichkeitsbewertung bestimmt. (vgl. Bild 5 und 6)

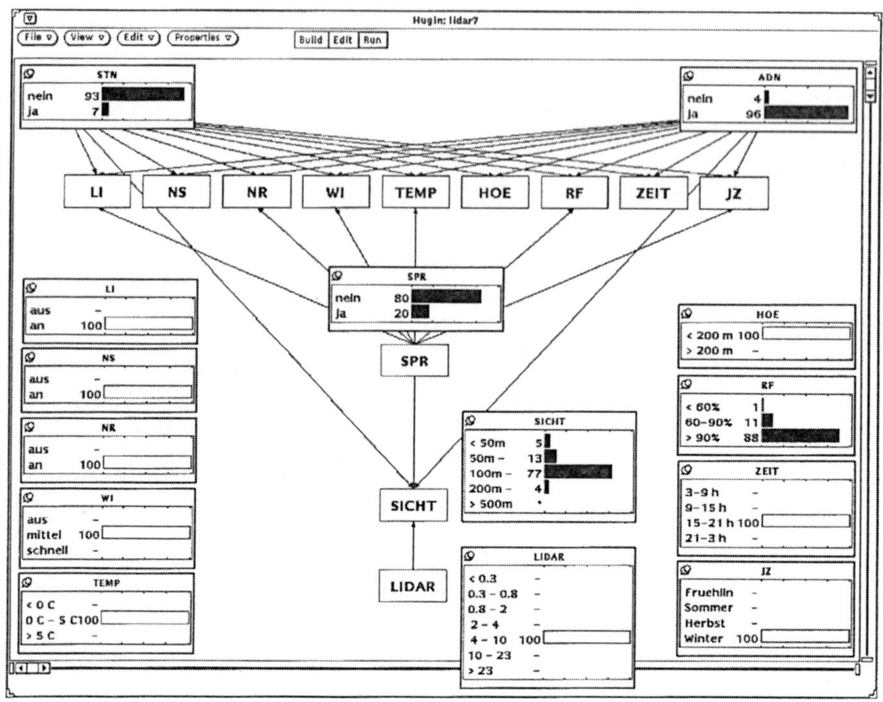

Bild 6: Eine LIDAR Messung für die Umweltsituation von Bild 5 möge einen Proportionalitätsfaktor für das Verhältnis von mehrfach zu einfach gestreutem Signalanteil von ca. 4 - 10 km^{-1} ergeben.

Eine Neubewertung des Netzes in Übereinstimmung zum bisherigen Wissen (dem alten Grundwissen und dem Zusatzwissen aus Bild 5) führt damit zu einer *neuen* Wahrscheinlichkeitsbewertung, welche ergibt, daß in diesem Beispiel mit "hoher" Wahrscheinlichkeit eine Normsichtweite von 100 - 200 m vorliegt.

Literaturverzeichnis

[1] Werner, Ch.; Oppel, U.; Streicher, J.; Krichbaumer, W.; Wallmeroth, K.; Findling, A.; Gatz, H.; Berghaus U.; Münkel, C.; Gelbke, E.: "LIDAR-Verfahren zur Bestimmung der Normsicht vom Fahrzeug aus", DLR, Oberpfaffenhofen, 1989.

[2] Werner, Ch.; Streicher, J.; Herrmann, H.; Dahn, H.G.: "Multiple-Scattering LIDAR Experiments", Optical Engineering 31 (8), p. 1731-1745, 1992.

[3] Berghaus, U.; Gelbke, E.; Herrmann, W.; Krichbaumer, W.; Münkel, Ch.; Oppel U.; Streicher, J.; Werner, Ch.: Patentschrift DE 39 30 272 C2 zu dem von der Bundesrepublik Deutschland am 25.7.91 erteilten Patent 39 30 272 mit der Bezeichnung "LIDAR zur Feststellung der Art und Abstandes von Trübungen in der Atmosphäre".

[4] Pearl, J.: "Probabilistic Reasoning in Intelligent Systems: Networks of Plausible Interference", Morgan Kaufmann, San Mateo, CA, USA, 1988.

[5] Oppel, U.: "HUGIN, a Software Package for the Construction of Expert Systems Based on Causal Probabilistic Networks", Proceedings of Workshop "Uncertainty in knowledge-based systems", Forschungsinstitut für anwenderorientierte Wissensverarbeitung (FAW) an der Universität Ulm, 8.-13.7.1990.

[6] Marchuk, G.I.; Mikhailov, G.A.; Nazaraliev, M.A.; Darbinjan, R.A.; Kargin, B.A.; Elepov, B.S.: "The Monte Carlo Methods in Atmospheric Optics", Springer, Berlin, 1980.

[7] Oppel, U.; Findling, A.; Krichbaumer W.; Krieglmeier, S.; Noormohammadian, M.: "A Stochastic Model for the Calculation of Multiply Scattered LIDAR Returns", DLR-FB 89-36, Köln, 1989.

[8] Oppel, U.; Starkov, A.; Noormohammadian, M.: "A Stochastic Model and a Variance Reduction Monte Carlo Method for the Calculation of Light Transport", Appl. Phy. B, erscheint 1993/94.

[9] Deirmendjian, D.: "Electromagnetic Scattering on Spherical Polydispersions", Elsevier Publishing Comp., New York, 1969.

[10] Shettle, E.P.; Fenn, R.W.: "Models for the Aerosols of the lower Atmosphere and the Effects of Humidity Variations on their optical Properties", Report AFGL-TR-79-0214, Airforce Geophysics Laboratory, Handscom, Mass., USA, 1979.

[11] Tries P.: "Sicht- und IR-Reichweite in der Atmospähre", Lehrgang S1.07, Carl-Cranz Gesellschaft, Oberpfaffenhofen, 13.-17.11.1989.

VDI/DIN-Working Group
VDI/DIN
Guidelines
for LIDAR
§ 1 ...
§ 2 ...
§ 3 ...

Standardisierung auf dem Gebiet der optischen Fernmeßverfahren in Deutschland

K. Weber
Fachhochschule Düsseldorf, FB 04, Josef-Gockeln-Str. 9, D - 40474 Düsseldorf

C. Weitkamp
GKSS - Forschungszentrum, Institut für Physik, Postfach 1160, D - 21502 Geesthacht

<u>Zusammenfassung</u>

Von der Kommission Reinhaltung der Luft (KRdL) im VDI und DIN wurde eine Arbeitsgruppe zur Standardisierung verschiedener Meßtechniken auf dem Gebiet der optischen Fernmeßverfahren gegründet. Die Arbeitsgruppe erarbeitet in verschiedenen Unterarbeitsgruppen Richtlinien/Normen für entsprechende Verfahren wie LIDAR, FTIR usw. Dieser Standardisierung kommt besonders im Hinblick auf eine Anwendung dieser Meßverfahren im Bereich der Umweltüberwachung eine hohe Bedeutung zu.

<u>Einführung</u>

Optische Fernmeßverfahren sind in den letzten Jahren für die verschiedensten Einsatzgebiete entwickelt und erprobt worden. Anwendungsfelder sind sowohl die Fernmessung von Luftverunreinigungen als auch verschiedener meteorologischer Parameter. Die Aktivitäten auf diesem Gebiet erstrecken sich gleichermaßen auf die Erforschung neuer Meßprinzipien und die Entwicklung von Meßsystemen wie auf die Durchführung von einzelnen Meßkampagnen und Langzeitmessungen mit diesen Systemen. Einen öberblick über die lebhaften Aktivitäten geben die zahlreichen Fachtagungen auf diesem Gebiet im nationalen und internationalen Bereich. Mittlerweile haben verschiedene Fernmeßsyteme einen technischen Stand erreicht, der sie aus dem reinen Wissenschafts- und Forschungsbereich heraushebt und kommerziell verfügbar macht. Damit werden diese Systeme auch für den Bereich der Umweltüberwachung und die Industrie interessant. Verschiedene potentielle Anwendungsgebiete im Bereich der Emissions- und Immisionsüberwachung sowie bei der Eigenüberwachung von Anlagenbetreibern wurden bereits veröffentlicht (siehe z.B. Stahl, Bröker und andere Autoren in [1] oder Hudson, Hunt und andere Autoren in [2]). Für diese Anwendungsfelder ist es jedoch wichtig, daß die Meßergebnisse einer hohen, feststellbaren und überprüfbaren Qualität genügen und gegebenenfalls auch juristisch belastbar sind. Darüberhinaus müssen Messungen, die von verschiedenen Personen oder mit verschiedenen Meßsystemen durchgeführt wurden, auch untereinander vergleichbar sein. Dies ist deswegen von großer Bedeutung, weil von Behörden angeordnete oder von diesen selbst durchgeführte Messungen unter Umständen gravierende Konsequenzen zur Folge haben können, wie z.B. Betriebs- oder Verkehrsbeschränkungen. Dies gilt in gleicher Strenge nicht im rein wissenschaftlichen Bereich, in dem die Qualität der Messungen eher einer Eigenverantwortung unterliegt.

Für die Qualitätssicherung bzw. Herstellung der Vergleichbarkeit der Messungen im Bereich der Umweltüberwachung und administrativer Meßaufgaben kommt der Standardisierung der Meßverfahren im Rahmen von Richtlinien und Normen eine große Bedeutung zu. Die Standardisierung kann sich dabei sowohl auf die Meßverfahren selbst als auch darauf beziehen, wie diese bei bestimmten Anwendungsfällen einzusetzen sind. In Deutschland wird die Überwachung der Luftqualität durch das Bundesimmissionschutzgesetz mit den zugehörigen Verordnungen und Verwaltungsvorschriften (z.B. TA Luft) geregelt. Gerade hier wird im Sinne einer Qualitätssicherung sehr häufig auf Standardisierungsdokumente wie z.B. VDI-Richtlinien zurückgegriffen, um einzelne Meßverfahren zur Messung von Luftverunreinigungen eindeutig festzulegen. Dies gilt bereits für punktförmig messende, schon etablierte Meßverfahren. In analoger Weise ist die Standardisierung von Fernmeßverfahren für den eben angesprochenen Bereich wichtig. Standardisierungsdokumente in diesem Sinne dienen darüberhinaus nicht nur der Qualitätssicherung, sondern erleichtern auch die Markteinführung von neuen Meßsystemen.

An die Kommission Reinhaltung der Luft im VDI und DIN wurde deshalb von verschiedener Seite der Wunsch herangetragen, Richtlinien bzw. Normen für unterschiedliche optische Fernmeßverfahren herauszugeben. Aus diesem Grunde wurde 1992 von der KRdL im VDI und DIN eine Arbeitsgruppe zu diesem Thema ins Leben gerufen.

Die Arbeit der Kommission Reinhaltung der Luft (KRdL) im VDI und DIN

Die Kommission Reinhaltung der Luft im VDI und DIN ist 1990 durch Zusammenschluß der vorher selbständigen Organisationseinheiten "VDI-Kommission Reinhaltung der Luft" (1957-1990) und dem "Normenausschuß NLuft im DIN" (1973-1990) entstanden. Dieses so entstandene Gemeinschaftsgremium ist ab 1990 für die originären Aufgaben beider Vorläufer-Organisationseinheiten zuständig. Die Arbeitsgebiete der KRdL umfassen technische und naturwissenschaftliche Bereiche der Luftreinhaltung einschließlich zugehöriger Grenzgebiete. Zu den Hauptaufgaben der KRdL im VDI und DIN gehören vor allem die Erstellung von

- VDI-Richtlinien
- DIN-Vornormen
- DIN-Normen
- DIN-EN-Normen
- DIN-ISO-Normen.

Weiterhin hat die KRdL im VDI und DIN die internationalen Sekretariate für das Technical Committee 146 "Air Quality" der Internationalen Organisation für Normung (ISO) sowie für das Technical Committee 246 des CEN (Europäisches Komitee für Normung) inne. In diesem Rahmen ist die KRdL im VDI und DIN auch für internationale und europäische Standardisierung im Bereich der Luftreinhaltung zuständig.

Darüberhinaus verfolgt und dokumentiert die KRdL den Stand der Technik auf diesem Gebiet in verschiedenen Veröffentlichungen, z.B der "Schriftenreihe der

KRdL", und ist Mitherausgeber der Zeitschrift "Staub - Reinhaltung der Luft". Außerdem werden von der KRdL im VDI und DIN zahlreiche Fachtagungen auf dem Gebiet der Luftreinhaltung durchgeführt.

<u>Organisatorisch und inhaltlich ist die KRdL in folgende Fachbereiche gegliedert:</u>

- Fachbereich I: "Entstehung und Verhütung von Emissionen"
- Fachbereich II: "Umweltmeteorologie"
- Fachbereich III: "Wirkungen von Staub und Gasen"
- Fachbereich IV: "Meßtechnik"
- Fachbereich V: "Verfahren zur Abgasreinigung, Staubtechnik"

<u>Richtlinien und Normen werden in allen diesen Bereichen erarbeitet.</u>

Von besonderer Bedeutung ist, daß die KRdL im VDI und DIN vom Umweltbundesamt institutionell gefördert wird. Aufgrund ihrer Nähe zur Rechtsordnung (Bundes-Immissionsschutzgesetz mit zugehörigen Verordnungen und Verwaltungsvorschriften) kommt den Richtlinien und Normen der Kommission Reinhaltung der Luft im VDI und DIN eine staatsentlastende Bedeutung zu.

Für die Richtlinien und Normen wird im allgemeinen von einer Arbeitsgruppe der KRdL, die sich aus Experten des jeweiligen Fachgebietes zusammensetzt, ein Entwurf erarbeitet. Dieser wird veröffentlicht und im Bundesanzeiger sowie in Fachzeitschriften angekündigt. Hierdurch ist Fachleuten und einschlägigen Institutionen die Möglichkeit des Einspruchs bzw. der fachlichen Einflußnahme möglich, bevor das Dokument endgültig als Richtlinie oder Norm verabschiedet wird. Aufgrund dieses Verabschiedungsverfahrens mit öffentlicher Einspruchmöglichkeit stellen die verabschiedeten Richtlinien und Normen nicht mehr nur das Arbeitsergebnis der einzelnen Experten der AG dar, sondern spiegeln den Stand der Technik und des Wissens aus einem weiten Bereich von Wirtschaft, Wissenschaft und Verwaltung wider. Dies unterstreicht die Bedeutung dieser Dokumente.

<u>Die Arbeit der Arbeitsgruppe "Fernmeßverfahren"</u>

Im Hinblick auf die vorstehenden Ausführungen kann es als sehr positiv angesehen werden, daß eine Arbeitgruppe "Fernmeßverfahren" innerhalb der KRdL im VDI und DIN gegründet wurde. Die Gründung wurde als Resultat eines Arbeitstreffens zum Thema "Fernmeßverfahren" am 19.5.92 in Karlsruhe, das von der KRdL einberufen wurde, von dort anwesenden Fachleuten einstimmig angeregt und danach von den entsprechenden Führungsgremien der KRdL beschlossen. Neben der Standardisierung der Fernmeßverfahren bietet diese AG auch ein Forum für Erfahrungsaustausch und Interessenabgleich zwischen Herstellern und Anwendern. Darüberhinaus bietet die AG den Vorteil, gemeinsame Standpunkte nach außen hin, auch im internationalen Bereich, deutlichzumachen.

Allen Beteiligten war von Anfang an klar, daß es nicht möglich ist, alle Fernmeßverfahren in einer einzigen Norm oder Richtlinie abzuhandeln, da die Spezifika der verschiedenen Fernmeßverfahren zu unterschiedlich sind. Aus diesem Grunde wurde beschlossen, verschiedene Richtlinien innerhalb einer Richtlinien-reihe herauszugeben. Außerdem kam man nach eingehender Diskussion zu der Entscheidung, die Normen / Richtlinien nicht nach Applikationen getrennt herauszugeben, sondern Standardisierungsdokumente zu den verschiedenen Meßtechniken wie FTIR, LIDAR usw. zu erarbeiten. Die einzelnen tandardisierungsdokumente sollten themenbezogen von entsprechenden Unterarbeitsgruppen erstellt und dann in der Gesamtgruppe diskutiert werden. Darüberhinaus sollte ein Glossar erarbeitet werden, in dem die besonderen Termini aus dem Bereich Fernmeßverfahren definiert und erläutert werden. Dies scheint für eine vereinheitlichte Sprachregelung für diese Meßverfahren zwischen Herstellern, Behörden, industriellen Nutzern und Forschungslabors geboten zu sein.

Entsprechend diesen Vorgaben hatte die AG Fernmeßverfahren sowie die Unterarbeitsgruppe für die FTIR-Verfahren bereits 1992 ihre Arbeit aufgenommen. Innerhalb der FTIR-Unterarbeitsgruppe wurde das oben schon erwähnte Glossar auf der Grundlage eines entsprechenden amerikanischen Dokumentes erstellt (siehe an anderer Stelle in diesem Band). Außerdem wurde ein erster Entwurf für ein Standardisierungsdokument "Fernmessung von Luftverunreinigungen mit der FTIR-Absorptionsspektroskopie bei offenen Absorptionsstrecken" in den wichtigsten Teilen erarbeitet. Dieses Dokument hat folgenden Inhalt:

<u>Vorbemerkung</u>

1. Einleitung
2. Grundlage des Verfahrens
3. Aufbau der Meßeinrichtung
4. Geräte und Betriebsmittel
5. Durchführen der Messung
6. Kalibrierung und Durchführung der Meßwertanzeige
7. Berechnen der Ergebnisse
8. Verfahrenskenngrößen und Einflußgrößen
9. Einsatzmöglichkeiten und Wartung

Anhang 1: Applikation Deponie
Anhang 2: Applikation KFZ-Abgase
Anhang 3: Applikation Industriestandort
Anhang 4: Screening

Während des internationalen Kongresses "Laser 93" fand am 23.6.1993 in München die konstituierende Sitzung der Unterarbeitsgruppe "LIDAR-Verfahren" statt. Die zunehmende Bedeutung der Fernmeßverfahren läßt sich daran ablesen, daß während dieser Sitzung Frau Dr. Gärtner von der Landesanstalt für Immissionsschutz in Essen ihrer überzeugung darüber Ausdruck gab, daß die neu gegründete AG u.a. einen wesentlichen Beitrag zur Standardisierung von Meßverfahren für die überwachung diffuser Quellen und für den mobilen Einsatz bei der Altlastensanierung erbringen könne. Im weiteren Verlauf der Sitzung wurde beschlossen, ein Standardisierungsdokument für die LIDAR-Verfahren zur Fernmessung von Luftverunreinigungen mit folgendem Inhalt zu erarbeiten:

<u>Vorbemerkung</u>

1. Einleitung
2. Grundlagen der LIDAR-Verfahren
   Rückstreu-LIDAR
   DAS-LIDAR (DIAL)
   Fluoreszenz-LIDAR
   Raman-LIDAR
   Doppler-LIDAR
3. Mathematische Beschreibung und Auswertung
4. Technische Implementierung
5. Kalibrierung und Durchführen von Messungen
6. Verfahrenskenngrößen
7. Übersicht über die Anwendbarkeit von LIDAR-Verfahren bei spezifischen Meßaufgaben
8. Vergleich mit Ergebnissen anderer Meßmethoden
9. Anwendungsbeispiele

Nach der konstituierenden Sitzung der LIDAR-Unterarbeitsgruppe wurde in der Gesamtarbeitsgruppe "Fernmeßverfahren" zusammen mit den Mitgliedern der FTIR-Unterarbeitsgruppe der Entwurf des Glossars diskutiert und überarbeitet (siehe Russwurm, Klein in diesem Band). Es wurde beschlossen, die AG "Fernmeßverfahren" bzw. die LIDAR- und FTIR-Untergruppen in regelmäßigem Turnus tagen zu lassen, um einen zügigen Fortschritt der Arbeiten zu erreichen. Eine Unterarbeitsgruppe zum DOAS-Verfahren soll zu einem späteren Zeitpunkt ebenfalls gegründet werden.

<u>Schlußbemerkung</u>

Innerhalb der Kommission Reinhaltung der Luft im VDI und DIN hat die Arbeitsgruppe "Fernmeßverfahren" mit ihren Unterarbeitsgruppen "LIDAR-Verfahren" und "FTIR-Verfahren" mit der Arbeit an Standardisierungsdokumenten begonnen. Ähnliche Standardisierungsarbeiten, insbesondere auf dem FTIR-Gebiet, wurden auch im Ausland von anderen Organisationen gestartet. Die Arbeiten werden in Kontakt mit den ausländischen Kollegen durchgeführt. Eine gemeinsame Standardisierung auf internationaler Ebene bei der ISO oder beim CEN ist zu einem späteren Zeitpunkt aufgrund der Struktur der KRdL im VDI und DIN durchaus denkbar und im Hinblick auf den Gemeinsamen Europäischen Markt auch sicherlich nützlich. Die Normen bzw. Richtlinien für die Fernmeßverfahren stellen ein wichtiges Element der Qualitätssicherung dar und werden einen Beitrag zur Vergleichbarkeit und Validierung derartiger Messungen liefern.

<u>Literatur:</u>

[1]   Werner, C., V. Klein, K. Weber (eds.); Laser in der Umweltmeßtechnik; Vorträge des 10. Internationalen Kongresses LASER 91; Springer-Verlag; Berlin, Heidelberg, New York; 1992
[2]   Proceedings of "Optical Remote Sensing, Applications to Environmental and Industrial Safety Problems"; A&WMA; Pittsburgh; 1992

Glossary of Terms for Optical Remote Sensing

George M. Russwurm
Man Tech Environmental Technology, Inc., P. O. Box 12313, Research Triangle Park,
N. C. 27709, USA

Volker H. Klein
Kayser-Threde GmbH, Wolfratshauser Str. 44-48, D - 81379 München

Remark: Words in CAPITAL LETTERS mark additional key words
within this glossary for further reference.

Absorption:

An interaction between matter (molecules of a trace gas) and electromagnetic radiation (light), which is one of the physical fundamentals for optical remote sensing systems. Absorption is deminishing electromagnetic radiation in a gas specific manner (LAMBERT BEER LAW). The absorption A is related to the TRANSMISSION T as follows:

$$A = 1 - T$$

Active System:

A optical remote sensing system that emits electromagnetic radiation (light) of an artificial source into the surrounding environment. A fraction of this radiation - gas specificly modified - is collected by its RECEIVER OPTICS and analyzed.

Average Concentration:

For FI-IR or DOAS systems this quantity is the result of dividing the integrated concentration (the quantity that is measured) by the path length used for the measurement. It has units of ppm, ppb, $\mu g/m^{**}3$, etc. (McClenny and Russwurm 1978)

Background Concentration:

The total amount of trace gases, which are widely abundant in the atmosphere due to natural or anthropogenic sources.

Background Spectrum:

With all other conditions being equal a spectrum taken in the absence of the particular absorbing species of interest.

Backscatter Lidar:

LIDAR system that investigates the properties of gaseous trace species or meteorological properties of the atmosphere by analyzing the backscattered fraction of emitted laser radiation (pulses);see also DIAL.

Band Pass Filter:

A filtering device that allows the transmission of only a narrow band of energies. The spectral width of this filter is characterized by its bandwidth.

Bandwidth:

The width of a spectral feature as recorded by a spectroscopic instrument.

Bistatic System:

A spatial arrangement in which the RECEIVER OPTICS (DETECTOR) is in some distance from the TRANSMITTER OPTICS (radiation source).

Blackbody:

RADIATION SOURCE that emits RADIATION in a spectral distribution close to the Planck law.

Broadband System:

Any system which analyzes an extended spectral range by its signal processing section.

Cooler:

A device used in infrared systems to cool the detector to achieve a low detection limit. There are two primary methods for cooling. One is to operate the cooler with liquid or gaseous nitrogen. The other is to use a closed cycle refrigerator (Stirling cooler). Higher detection limits at lower costs can be achieved by using e.g. electrically operated coolers (Peltier elements).

Correlation Spectrometer:

Instrument for a gas specific investigation that correlates spectral signatures of the species to be analyzed with reference spectra. One group of instruments uses REF-

ERENCE SPECTRA that are stored within the instrument by means of glass disks with etched spectral features. Another group uses reference vials to generate reference spectra. Both groups of instruments can be operated in the passive and in the active mode.

Differential Absorption Spectroscopy (DAS):

DAS uses two frequencies emitted by the same laser or by different lasers to perform measurements of the concentration of gas along a given line of sight. The frequency of one laser line (signal frequency) is tuned to the frequency of the line centre of the absorption feature and the frequency of the other laser line (reference frequency) is tuned besides this feature. The difference of the amount of transmitted light at these two frequencies is the quantity used for the measurement.

Detector:

Radiation sensitive element that converts electromagnetic RADIATION into an electric signal. A detector is characterized by its spectral response (sensitivity as a function of wavelength or frequency).

DIAL:

An acronym for DIfferential Absorption Lidar. This technique is based on the combination of a BACKSCATTER LIDAR and DIFFERENTIAL ABSORPTION SPECTROSCOPY (DAS). A DIAL system uses two pulsed frequencies emitted by the same laser or by different lasers to perform range resolved measurements of the concentration of gas along a given line of sight. The frequency of one laser line (signal frequency) is tuned to the frequency of the line centre of the absorption feature and the frequency of the other laser line (reference frequency) is tuned besides this feature. The difference of the amount of backscattered light at these two frequencies is the quantity used for the measurement. The analysis of time between the emission of the laser pulses and the detection of the backscattered signal provides the range resolution of the measurement.

DOAS:

An acronym for Differential Absorption Spectroscopy. A technique whereby, in principle, any know difference in absorbency is used to determine the concentration of a gas. Generally the absorption difference is taken between the spectral line centre and the wing.

Electromagnetic Spectrum:

The total of all possible frequencies of electromagnetic radiation.

Extinction:

The summerized effect of the two diminishing effects ABSORPTION and SCATTER-ING.

Fingerprint Region:

The region of the absorbency spectrum of a trace gas that essentially allows the un-equivocal identification of this particular gas. For optical remote sensing this region is normally covering the wavenumber range from 650 to 1300 wavenumbers. (Willard et al. 1974)

Flux:

The amount of particles (mass flux) or photons (radiation flux), passing a given unit area of surface per unit of time (Calvert 1990).

Fourier-Transformation:

A mathematical operation to convert an aperiodic function into an integral sum over a continuous range of frequencies (Champeney 1973). The Fourier transformation of an interferogram of a Michelson interferometer within a FT-IR system converts the record-ed intensity (function of the optical path length) to a spectral signal (function of wavenumber).

Free Atmosphere:

Part of the atmosphere that is exclusively characterized by actual meteorological con-ditions.

Infrared Spectral Range:

That portion of the ELECTROMAGNETIC SPECTRUM that covers the region from about 10 wavenumbers to about 13000 wavenumbers. It is divided into three regions

(DIN 5031/7):

1. The near infrared (from 13000 to 3300 wavenumbers).
2. The mid infrared (from 3300 to 200 wavenumbers).
3. The far infrared (from 200 to 10 wavenumbers).

Interference:

Those physical effects caused by superimposing two or more light waves. The princi-pal of superposition states that the total amplitude of the electromagnetic disturbance at a point is the vector sum of the individual electromagnetic components incident there.

Interferogram:

The observed or recorded effects of INTERFERENCE. The unprocessed output of a FT-IR instrument is an interferogram and is the primary data to be collected and stored (Stone 1963; Griffiths and de Haseth 1986).

Interferometer:

Any of several kinds of instruments used to produce interference effects. The Michelson interferometer used in FT-IR instruments is the most famous of a class of interferometers that produce INTERFERENCE by division of wavefront (Toansky 1962).

Lambert-Beer Law:

The Lambert Beer law states that the intensity of a monochromatic plane wave incident on an absorbing medium of constant thickness diminishes exponentially with the number of absorbers in the beam. Strictly speaking Beer's law holds only of the following conditions are met:

1: Perfectly monochromatic radiation.
2: There is no scattering.
3. The beam is strictly parallel.
4. The absorbing molecules are never close enough to one

another or other molecules that the energy levels are affected (Pfeiffer and Liebhafsky 1951; Lothian 1963). For an excellent discussion of the derivation of the Lambert-Beer law see Penner (1959).

Laser:

An acronym for the term Light Amplification by Stimulated Emission of Radiation. A source of light that is highly coherent both spatially and temporally (Lengyel 1971) and is emitted on one or more wavelengths. There are two groups of lasers which are operating either in a pulsed mode or continuously. Most of the active optical remote sensing systems are using lasers as light source.

Lidar:

An acronym for the term LIght Detection And Ranging. This group of monostatic active remote sensing systems detects the presence of TRACE GASES or analyzes meteorological parameters by measuring the backscattered portion of a laser beam. Lidar systems determine range by electronically gating the detected signal and using the fact that LIGHT travels about 1 meter per 3 nanoseconds (Calvert 1990).

Light:

Strictly light is defined as that portion of the ELECTROMAGNETIC SPECTRUM that causes the sensation of vision. It extends from about 25.000 wavenumbers to about 14,300 wavenumbers (Halliday and Resnick 1974)

Light Scattering:

The redirection of light waves due to interaction with molecules or aerosols. If the size of the body causing the SCATTERING is small compared to the wavelength of the incident radiation (e.g. molecules) the SCATTERING is termed Rayleigh scattering. If the size is large compared to the wavelength (e.g. aerosols) the scattering is termed Mie scattering. As light travels through a medium, two physical processes diminish the intensity in the forward direction; SCATTERING and ABSORPTION. The sum of these two effects is calls EXTINCTION (van de Hulst 1981).

Long Path Monitoring:

A monitoring technique that uses a long, extended path. Laser based DAS systems can make measurements over a path length of a few kilometers. DOAS system can make measurements of ozone using a path up to 2 kilometers. FT-IR systems generally use path of less than 1 kilometer.

Minimum Detectable Concentration:

The minimum concentration of a compound that can be distinguished from zero signal with a given statistical probability of 95 % (VDI 2449/2).

Monostatic Arrangement:

A remote sensing system with the RADIATION SOURCE and the RECEIVER OPTICS at the same end of the path. For FT-IR systems and long path monitoring systems the beam is generally returned by a RETROREFLECTOR. For LIDAR systems the back scattered portion of the laser beam is measured directly.

Open Ended Arrangement:

A remote sensing system in which the remote sensor uses LIGHT reflected from targets of opportunity (walls, trees, etc.) or skylight as a SOURCE.

Open Path Monitoring:

A type of remote sensing that uses a path that is completely open to the atmosphere. Thus changes in concentration of a particular gas are transported into the beam by winds or diffusion or both. The open path is the most frequently used in remote sensing.

Optical Depth:

The negative logarithm of the extinction [ln I/Io], where I is the intensity of the LIGHT at the back plane of the absorbing medium and Io is the incident intensity. The optical depth is the product of the extinction coefficiant, the density of the gas and the length of the transmitted gas layer (LAMBERT BEER LAW).

Optical Remote Sensing:

A generic term used to describe any of a number of optical measurement techniques that measure the quantity of atmospheric constituents across an extended range. These techniques include DIAL, DOAS, FT-IR, GASPEC, LIDAR etc. One feature common to all these techniques is that they do not collect a sample that has to be analyzed.

Parts Per Million Meters (ppm x m):

The units term for the quantity that is measured by many remote sensors. It is the unit associated with the quantity PATH INTEGRATED CONCENTRATION. Is a possible unit of choice for reporting data from remote sensors since it independent of the path length.

Passive System:

Any optical remote sensing system with a spectroscopic analyzing detector that does not radiate energy into its surroundings.

Path Integrated Concentration:

The quantity that is measured by a remote sensor using a long path. I has units of concentration times length, e.g. ppm x m or ppb x km.

Path Averaged Concentration:

The result of dividing the path integrated concentration by the path length. It gives the average value of the concentration along the path (McClenny and Russwurm 1978).

Plume:

The gaseous and aerosol effluents emitted from a chimney or other source and the volume of space they occupy. The shape of a plume and the concentration of pollutants within are very sensitive to meteorological conditions (Calvert 1990).

Radiation:

Electromagnetic energy that is used to detect gaseous trace components in the atmosphere. For optical remote sensing systems the frequency of this radiation is limited to the regions of UV, VIS or IR light.

Radiation Source:

See SOURCE.

Radiometry:

The measurement of various quantities such as intensity associated with radiant energy (Calvert 1990). This is in contrast to the term photometry which assumes spectral sensitivity of the human eye as the DETECTOR (Walsh 1965)

Real Time System:

A monitoring system that acquires and records data at a rate that is at least comparable to the rate at which the concentration is changing.

Receiver Optics:

Arrangement of optical elements such as telescopes, mirrors or lenses to collect electromagnetic radiation for a given direction and to focus this optical signal onto the analyzing unit (e.g. DETECTOR).

Reference Spectrum:

Spectrum of the absorbance versus wavenumber for a well known pure sample or a set of gases. The spectra are obtained under controlled conditions of pressure and temperature and with known concentrations. For most instruments the pure sample is pressure broadened with nitrogen so that the spectra are representative for atmospherically broadened lines. These spectra are used for obtaining the unknown concentrations of gases in ambient atmospheric samples (total pressure appr. 1 bar and temperature appr. 300 K). Reference spectra for the determination of absolute concentrations of unknown gases are usually recorded in form of data bases.

Resolution:

The minimum separation that two spectral features can have and still, in some manner, can be distinguished from one another. A common requirement for two spectral features to just be considered resolved is the Rayleigh criterion. This states that two features are just resolved when the maximum intensity of one falls at the first minimum of the other (Tolansky 1962).

Retroreflector:

The CIE (Commission Internationale de L'Eclairage) defines retroreflection as radiation returned in directions close to the direction from which it came, this property being maintained over wide variations of the direction of the incident radiation. Retroreflector devices come in a variety of forms and have many uses. The one commonly described by workers in remote sensing uses total internal reflection from three mutually perpendicular surfaches. This kind of retroreflector is usually called a corner cube of prismatic retroreflector (Rennilson 1980).

Scattering:
See LIGHT SCATTERING

Source:

The device that supplies the electromagnetic energy for the varios instruments used to measure atmospheric gases. These are generally a Nernst glower or glow bar for the infrared region or a Xenon arc lamp for the ultraviolet region. Also there are several types of LASERS that are used as sources for DIAL and LIDAR instruments. See also RADIATION SOURCE.

Spectral Interference:

When the absorbance features from two or more gases cover the same wavenumber regions the gases are said to exhibit spectral INTERFERENCE. Water vapor produces the strongest spectral interference for infrared spectroscopic instruments.

Spectral Signature:

Specific modulation of electromagnetic radiation due to gaseous compounds that allows the identification of these gases by spectroscopic methods

Spectrum:

Intensity distribution of electromagnetic radiation as a function of wavelength, WAVENUMBER or frequency.

Trace Gas:

Gaseous atmospheric compound with concentration well below the concentration of the major compounds.

Transmission:

Fraction of radiation that is not absorbed or scattered by a layer of gas. Transmission T is related to ABSORPTION A by the expression

$$T = 1 - A$$

Transmitter Optics:

Part of an active remote sensing system that collimates the emitted radiation to a desired direction. Transmitter optics also reduce the effect of beam divergence within LASER based ystems.

Unistatic System:

A system that has the source and the receiver at the same place.

Wave Number:

The number of waves per centimeter. This term has units of $cm^{**}-1$.

References:

Calvert, J. G.; Glossary of Atmospheric Chemistry terms (Recommendations 1990) Pure and Appl Chem., 62, 2167, 1990

Champeney, D. C.; Fourier Transforms and their Applications; Academic Press, 1973

Griffiths, P. R. and de Haseth, Fourier Transform Infrared Spectromety, J. A. Wiley and Sons, 1986

Halliday, D. and Resnick R.; Fundamentals of Optics, McGraw-Hill, 1950

Lengyel, B. A.; Lasers, 2nd Ed., Wiley-Intersience, 1971

McClenny W. A. and Russwurm, G. M.; Laser-Based Long Path Monitoring of Ambient Gases-Analysis of Two Systems; Atmos. Environ., 12 1443, 1978

Penner, S. S.; Quantitative Molcular Spectroscopy and Gas Emissivities; Addison-Wesley, 1959

Pfeiffer, H. G. and Liebfahsky, H. A.; The Origins of Beer's Law; J Chem Educ., 28, 123, 1951

Rennilson, J. J.; Retroreflection measurements: A Review; Applied Optics, 19, 1234, 1980

Stone, J.M.; Radiation and Optics; McGraw-Hill, 1963 Tolansky, S.; An Introduction to Interferometry.; John Wiley and Sons, 1962

van de Hulst, H. C.; Light Scattering by Small Particles; Dover Publications, 1981

Walsh, J. W. T.; Photometry; Dover Publications, 1965

Willard, H. H., Merritt, L. L., Dean, J. A.; Instrumental Methods of Analysis, 5th Ed.; D. Van Nostrand, 1974